Health and Safety at Work: Key Terms

Health and Safety at Work: Key Terms

Jeremy Stranks
MSc, FCIEH, FIOSH, RSP

OXFORD AMSTERDAM BOSTON LONDON NEW YORK PARIS
SAN DIEGO SAN FRANCISCO SINGAPORE SYDNEY TOKYO

Butterworth-Heinemann
An imprint of Elsevier Science
Linacre House, Jordan Hill, Oxford OX2 8DP
225 Wildwood Avenue, Woburn, MA 01801-2041

First published 2002

British Library Cataloguing in Publication Data
A catalogue record for this book is available from the British Library

Library of Congress Cataloguing in Publication Data
A catalogue record for this book is available from the Library of Congress

ISBN 0 7506 5446 5

For more information on all Butterworth-Heinemann publications visit our website at www.bh.com

Typeset at Replika Press Pvt Ltd, Delhi 110 040, India
Printed and bound in Great Britain by Biddles Ltd www.biddles.co.uk

Preface

Occupational health and safety is a diverse subject embracing many disciplines, such as engineering, law, occupational psychology, construction, physics and chemistry. On this basis, both health and safety practitioners and those studying for qualifications in the subject may encounter terms with which they are unfamiliar.

Health and Safety at Work: Key Terms has been written principally as a revision aid for people studying for examinations in the subject, in particular the examinations set by the National Examination Board in Occupational Safety and Health (NEBOSH). To this extent, in the identification and selection of key terms, the current syllabus for the NEBOSH National Diploma in Occupational Safety and Health was taken into account and the syllabus reference is incorporated where appropriate.

This book is also targeted at those who may have a limited involvement in health and safety, such as lawyers, human resources managers, engineers and those involved in construction activities.

I hope that all those who use this book will find it helpful.

Jeremy Stranks

A

Abatement Notice Where a local authority is satisfied that a statutory nuisance exists, or is likely to occur or recur, the local authority is empowered to serve an Abatement Notice, imposing all or any of the following requirements:

- requiring the abatement of the nuisance or prohibiting or restricting its occurrence or recurrence;
- requiring the execution of any works, and the taking of other steps, that may be necessary for any of these purposes;

and the notice shall specify the time or times within which the requirements of the notice are to be complied with.
(Environmental Protection Act 1990)

ability The potential for the acquisition of skills.

abrasion A form of wear in machinery whereby small and hard particles are entrapped between moving surfaces, resulting in scoring of the surfaces.

absolute liability (1B1) Certain duties under health and safety law are of an absolute nature, and are not qualified by the terms 'so far as is practicable' or 'so far as is reasonably practicable'.
Absolute duties are qualified by the term 'shall' or 'must', e.g. the absolute duty to maintain work equipment in an efficient state, in efficient working order and in good repair under the Provision and Use of Work Equipment Regulations 1998.
When charged with an absolute offence it may be possible to submit a plea in mitigation.

absolute threshold The smallest amount of a stimulus that can be perceived.

absorbed dose (radiation) The unit of energy deposited in the body as a result of radiation, expressed in grays (Gy); 1 gray is equivalent to 1 joule per kilogramme.

absorption The entry of a substance into the body. This may be by inhalation, pervasion (through the skin), ingestion, injection, inoculation and implantation.

absorption (sound) A form of airborne noise control using porous lightweight barriers frequently mounted on walls or suspended from ceilings.

accelerometer (2E2) An instrument for the measurement of vibration. The transducer mainly used for vibration measurement is the piezoelectric accelerometer, which exhibits better all round characteristics than other types of vibration transducer.

It has a very wide frequency and dynamic ranges with good linearity throughout the ranges.

access equipment (1C2) There is an absolute and general duty on employers, the self-employed and controllers to ensure safe means of access to and from any place of work, which may be above or below ground. Any access equipment must be erected under the supervision of a competent person.
Access equipment may include scaffolds, ladders, roof ladders, lightweight scaffolds, access towers, safety harnesses, safety nets and lift trucks incorporating a working platform.
[Construction (Health, Safety and Welfare) Regulations 1996]

accessory for lifting Work equipment for attaching loads to machinery for lifting. (Lifting Operations and Lifting Equipment Regulations 1998)

access to information Where an employee is suing an employer, he or she will need access to certain information.
Under section 28 of the HSWA, as amended by the Employment Protection Act 1975, anyone involved in civil proceedings may obtain a written statement of relevant factual information obtained in the exercise of his or her powers from an inspector.
Under section 27 of the HSWA, the Health and Safety Commission (HSC) can require any person to provide specific information concerning health and safety matters, either to the HSC or to an enforcing authority.

Accident Book Employers must provide an Accident Book (Form BI 510) in which employees must enter the 'appropriate particulars' of all accidents. Where an employee is incapacitated, the entry must be made by another employee. Accident Books, when completed, must be retained by the employer for 3 years after the date of the last entry.
(Social Security Act 1975)

accident causation (2A1) Various theories as to the causes of accidents have been proposed over the last 70 years. These include:

- the Pure Chance Theory;
- the Biased Liability Theory;
- the Accident Proneness (Unequal Initial Liability) Theory;
- the Theory of Unconscious Motivation;
- the Adjustment Stress Theory;
- the Goals–Freedom–Alertness Theory;
- the Domino Theories;
- the Epidemiological Approach;
- the Modelling Approach;
- the ILCI Loss Causation Model;
- the Multiple Causation Theory;
- the Cause–Accident–Result Sequence.

See individual entries.

accident costs All accidents and occupational ill health represent some form of cost to an organization.

Direct costs of accidents include:

- claims made in a civil court;
- fines imposed in criminal courts; and
- increased employer's liability insurance premiums.

Indirect costs of accidents include:

- medical and first aid treatment;
- transport to hospital;
- lost time of the injured employee;
- loss of an injured person's personal skills;
- lost time of other persons – managers, supervisors and employees;
- replacement labour, including training, retraining and extra supervision;
- lost production;
- repair or replacement of damaged plant, equipment, intermediate and finished products;
- health and safety administration and investigation;
- legal costs – court preparation, legal fees, witness and expert witness fees;
- loss of employee morale, resulting in reduced productivity;
- counter-productive trade union activity.

accident (definitions) The term 'accident' has been defined thus:

- An unforeseeable event often resulting in injury (Oxford Dictionary).
- An event that is without apparent cause or unexpected; unintentional act, chance, misfortune; including event, especially one causing injury or damage (Concise Oxford Dictionary).
- Any deviation from the normal, the expected and the planned, usually resulting in injury (Royal Society for the Prevention of Accidents).
- A management error; the results of errors or omissions on the part of management (British Safety Council).
- An uncontrolled and unplanned event in which the action or reaction of an object, substance, person or radiation results in personal injury or property damage (H. W. Heinrich).
- An unintended or unplanned happening that may or may not result in personal injury, property damage, work process stoppage or interference, or any combination of these conditions under such circumstances that personal injury might have resulted (Frank Bird, American exponent of Total Loss Control).
- An undesired event that results in physical harm or damage to property. It is usually the result of contact with a source of energy above the threshold limit of the body or structure.
- An unexpected, unplanned event in a sequence of events that occurs through a combination of causes, resulting in physical harm (injury or disease) to an

individual, damage to property, business interruption or any combination of these effects (Health and Safety Unit, University of Aston in Birmingham).

- An unexpected, unplanned occurrence which may involve injury (International Labour Office).

accident proneness The concept or notion that some people are more liable to have accidents than others or that some people are more susceptible to accidents than others.

Accident Proneness (Unequal Initial Liability) Theory A widely discussed theory in accident causation research, this theory proposes that there exists a certain sub-group of the general population that is more liable to have accidents. The theory is based on certain innate personality characteristics that cause accident prone individuals to have more accidents than people who are not accident prone.

accident ratios (1A4) Studies of accidents resulting in death, major injury, minor injury, lost time, property damage and 'near misses' over the last 40 years have produced a number of accident ratios. The principal objective of these studies was to establish a relationship between 'near misses' and other categories of accident.

The philosophy behind the use of accident ratios is that by investigation of, and taking action to prevent, 'near misses', the more serious types of accident can be avoided.

The various accident ratios are shown in Figure A.1.

accident statistics (1A5) Accident statistics, as a form of safety monitoring, generally take the form of a number of standard indices:

$$\text{Frequency rate} = \frac{\text{Total number of accidents}}{\text{Total number of man hours worked}} \times 100\,000$$

$$\text{Incidence rate} = \frac{\text{Total number of accidents}}{\text{Number of persons employed}} \times 1000$$

$$\text{Severity rate} = \frac{\text{Total number of days lost}}{\text{Total number of man hours worked}} \times 1000$$

$$\text{Mean duration rate} = \frac{\text{Total number of days lost}}{\text{Total number of accidents}}$$

$$\text{Duration rate} = \frac{\text{Number of man hours worked}}{\text{Total number of accidents}}$$

Accident incidence rate is the most commonly used index.

achievement motivation One of the theories of risk taking developed by McClelland and, subsequently, Atkinson, and derived from motivation theory.

McClelland studied business risk taking, indicating that 'doing your own thing', such as establishing a business, was a strong motivator for some people, compared with those who preferred the relative security of working for a large organization.

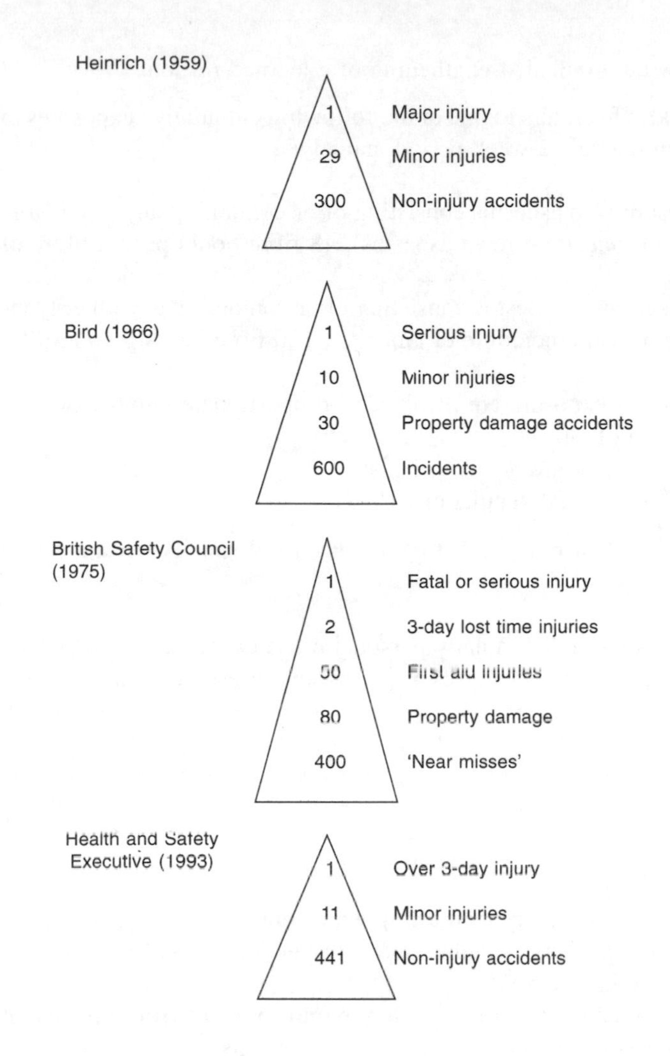

Figure A.1 Accident ratios.

This achievement motivated these business risk takers to greater efforts, and they actually enjoyed the financial risk taking that this entailed.
Atkinson suggested that there are various motivational factors affecting people, in particular the fear of failure (negative) and the more positive motivation towards success, which can be measured by a well-validated questionnaire.

achievement need A need to succeed and strive for standards of excellence; something which serves to motivate an individual to do well.

acoustic trauma A condition involving sudden aural damage resulting from short-term intense exposure to noise, or even from one single exposure, such as gunfire or explosion.

acquisition The slow but gradual strengthening of a learned response.

action level (asbestos) This refers to one of the following cumulative exposures to asbestos over a continuous 12-week period, namely:

- where the exposure is to asbestos consisting of or containing any *crocidolite* (blue asbestos) or *amosite* (brown asbestos), 48 fibre hours per millilitre of air;
- where the exposure is to asbestos consisting of or containing any other types of asbestos (other than crocidolite or amosite), 120 fibre hours per millilitre of air;
- where both types of exposure are involved, a proportionate number of fibre hours per millilitre of air.

(Control of Asbestos at Work Regulations 1987)

action level (noise) The Noise at Work Regulations 1989 specify certain action levels as follows:

- the first action level – means a daily personal noise exposure of 85 dB(A);
- the peak action level – means a level of peak sound pressure of 200 pascals;
- the second action level – means a daily personal noise exposure of 90 dB(A).

Every employer must, when any of his employees is likely to be exposed to the first action level or above, or to the peak action level or above, ensure that a competent person makes a noise assessment.
(Noise at Work Regulations 1989)

active monitoring A form of safety monitoring which entails a range of exercises directed at preventing accidents, including safety inspections, safety audits, safety tours and safety sampling exercises.
The undertaking of a range of proactive safety monitoring activities directed at preventing accidents, such as safety audits and inspections.

activity rate The rate at which spontaneous decay takes place in a radionuclide. Activity rate is expressed in becquerels (Bq); 1 becquerel corresponds to the decay of 1 radionuclide per second.

Act of Parliament (Statute) (1B2) The principal form of legislation which has gone through the parliamentary process to become a statute, e.g. the Health and Safety at Work etc. Act (HSWA) 1974.
The HSWA enables the Secretary of State for Employment to make Regulations on a range of health and safety related matters, e.g. the Control of Substances Hazardous to Health Regulations 1999.

actual dismissal The situation where an employer terminates a contact of employment.

acute effect (1E1) A rapidly produced effect on the body following a single exposure to an offending or hazardous agent.

adaptation A change in the sensitivity of a sense organ due to stimulation or lack of stimulation.

adjustable guard A machinery guard incorporating an adjustable element which, once adjusted, remains in that position during a particular operation. See Figure A.2.
(BS 5304: *Safeguarding of Machinery*)

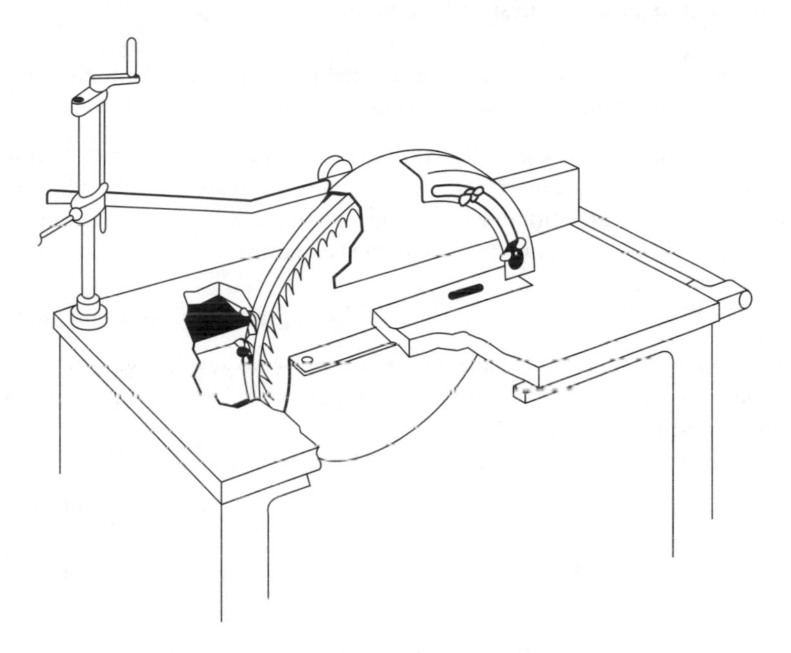

Figure A.2 Adjustable guards to a floor-mounted circular saw.

adjustment The relationship that exists between a person and his environment in the satisfaction of his motives.

Adjustment Stress Theory A theory of accident causation developed by Kerr, which states that individuals who cannot reach some form of adjustment with their work and working environment will tend to have more accidents than others. This failure to adjust is caused by a range of physical and psychological stressors.

Advisory Committee on Toxic Substances (ACTS) (1E2, 2E5) A formally constituted committee operating under the auspices of the Health and Safety Commission (HSC) and including representatives from the Confederation of British Industry (CBI), the Trades Union Congress (TUC), local authorities, government departments and independent experts. This committee provides advice to the HSC on the toxicological aspects of existing and new substances used at work.

Advisory, Conciliation and Arbitration Service (ACAS) ACAS was established by the Employment Protection Act 1975 and has the following functions:

- to advise employers and trade unions on any matter concerned with employment policies and industrial relations;
- to conciliate in matters such as maternity leave, trade union membership and activities, dismissal and redundancies;
- to arbitrate at the request of both parties.

aerosol Any combination of particles carried in, or contained in, air. An aerosol may embrace liquid droplets as well as solid particles.

aerosol dispenser An article which consists of a non-reusable receptacle containing a gas compressed, liquefied or dissolved under pressure, with or without liquid, paste or powder and fitted with a release device allowing the contents to be ejected as solid or liquid particles in suspension in a gas, as a foam, paste or powder, or in a liquid state.
[Chemicals (Hazard Information and Packaging for Supply) Regulations 1994]

agent In relation to any client means any person who acts as agent for a client in connection with the carrying on by the person of a trade, business or other undertaking (whether for profit or not).
[Construction (Design and Management) Regulations 1994]

airborne sampler A device for sampling both dust and radiation levels whereby a known volume of air is drawn through a filter over a specific sampling period. In the case of dust, this is measured gravimetrically, whereas with radiation, the filter paper is scanned, using a counter.

air flow This is an important factor in the assessment of comfort, air movement being perceptible at around 9 m/min and extremely uncomfortable above 30 m/min. The sensation of air flow is directly related to air temperature and skin sensitivity.

air line breathing apparatus A form of respiratory protective equipment comprising a full or half face mask connected via a flexible hose either to a source of uncontaminated air, in the case of short distances, or to a compressed air line via a filter and demand valve.

air receiver Any vessel (other than a pipe or coil, or an accessory, fitting or part of a compressor) for containing compressed air and connected with an air compressing plant; or
any fixed vessel for containing compressed air or compressed exhaust gases and used for the purpose of starting an internal combustion engine; or
any fixed or portable vessel (not being part of a spraying pistol) used for the purposes of spraying by means of compressed air any paint, varnish, lacquer or similar material; or
any vessel in which oil is stored and from which it is forced by compressed air.
(Factories Act 1961)

air sampling The process of taking a sample of air for subsequent analysis. It may be undertaken on a short-term or long-term basis.
Short-term sampling (grab sampling, snap sampling) implies taking an immediate sample of air and, in most cases, passing it through a particular chemical reagent which responds to the contaminant being monitored.
Long-term sampling can be undertaken using personal sampling instruments or dosemeters, which are attached to the individual, and by the use of static sampling equipment located in the working area.

alarm reaction The first stage of the General Adaptation Syndrome in which a person reacts vigorously to a stressful situation (Selye).

all reasonable precautions and all due diligence A defence available to a person charged with an offence under certain regulations, such as the Control of Substances Hazardous to Health Regulations 1999, and in some civil actions. To rely on this defence, an employer must establish that, on the balance of probabilities, he has taken *all* reasonable precautions and exercised *all* due diligence to avoid the commission of the offence.

alpha particles A form of ionizing radiation consisting of two protons and two neutrons bound together.
Alpha particles, which are heavy and double charged, are, fundamentally, helium atoms that have lost their two orbiting electrons and which are positively charged.

alternating current An electric current where the direction of flow alters at regular intervals.

alternative means of escape A second route, usually in the opposite direction to the first means of escape, but which may join the first means of escape from a building.

amplitude (2E2) The amplitude of a sound wave is the maximum displacement of a particle from its rest position. The amplitude of a sound wave determines loudness, although the two are not directly related.

analogy An epidemiological term indicating the properties of certain substances or agents which may create comparable or similar effects on health.

annealing The process of heating a metal or alloy to some predetermined temperature below its melting point, maintaining that temperature for a time, and then cooling slowly.

anthropometry (2D2) The study and measurement of body dimensions, the orderly treatment of the resulting data, and the application of the data in the design of workspace layouts and equipment.

anti-priming pipe A pipe fitted in a boiler to prevent the loss of water carried over from one part of the boiler to another part as a result of siphonic action.

anxiety A state of tension coupled with apprehension, worry, guilt, insecurity and a constant need for reassurance.
A vague or objectless fear.

appeals procedure – civil courts The Court of Appeal (Civil Division) and the House of Lords are the courts of appeal in civil cases, hearing appeals from the lower courts, i.e. the County Courts and the High Court.

appeals procedure – criminal courts Where convicted of a summary offence, there is a right of appeal to a Crown Court against:

- conviction, sentence or both, if the accused pleaded not guilty;
- sentence only if he pleaded guilty.

The prosecution has no right of appeal.
Appeals from the Crown Court are heard by the Court of Appeal (Criminal Division).

Approved Code of Practice (1B2) Section 16 of the HSWA gives the HSC power to prepare and approve codes of practice (ACOPs) on matters contained not only in regulations, but in sections 2–7 of the HSWA.
The HSE, acting on behalf of the HSC, must consult with any interested body prior to the HSC approving a code.
An ACOP is a quasi-legal document; non-compliance does not constitute a breach of the law. Failure to comply with an ACOP may be taken by a Court in criminal proceedings as proof that a person has contravened the regulation or sections of the HSWA to which the provision relates.
In the above situation, it will be open to the person to satisfy a Court that he or she has complied with the regulation or section in some other way, e.g. works of equivalent nature.

Approved Supply List (1E1) The list entitled 'Information Approved for the Classification and Labelling of Substances and Preparations Dangerous for Supply' approved by the HSC, comprising Parts I–IV, together with such notes and explanatory material as are required for the use of the List.
The Approved Supply List is issued in conjunction with the Chemicals (Hazard Information and Packaging for Supply) Regulations 1994.

aptitude (1A2) The potential that an individual has to profit from a certain type of training. It indicates how well that person would be able to undertake a task after receiving the training.

arcing A form of electrical discharge which occurs when the voltage in a conductor is great enough to create a path between that conductor and another conductor which is at lower voltage. The arc created is capable of crossing or 'jumping' the air gap or insulation which separates the two conductors, significant amounts of energy being created in a short time.
The process of arcing generates ultraviolet radiation, which can burn the retina of the eye in particular, together with the skin. Arcing may also be a source of ignition.

arousal An increase in alertness and muscular tension. It may be associated with excitement, physical exercise, fear, expectation and certain stressful events.

arousal theory One of the theories of risk taking based on the fact that, whilst most people avoid taking risks, there is a small proportion of the population who actually enjoy risk taking, including, for instance, compulsive gamblers and certain drivers. This relationship between arousal and personal risk taking is based on the fact that when people take risks it raises their level of arousal to a point where they reach a 'high'. Once that 'high' is reached, the level of risk taking reduces.
Studies by F. W. Taylor examined the relationship between arousal and galvanic skin response (the basis for the lie detector test) in driving situations. It was noted that, as the road conditions deteriorated, so the level of arousal increased.

Art of Plain Talk In a particular study into human understanding carried out by Rudolf Flesch (1946), Flesch analysed the elements of language expression that, generally, make for ease of comprehension.
These are:

- The number of words in a sentence; the shorter the sentence, the more easily it will be understood.
- The number of syllables; the shorter the length of the words used, the easier they are understood.
- The number of personal words and sentences; the greater the number or percentage of personal words and/or sentences, the more comprehensive the language.

asbestos A composition of the silicates of iron, magnesium, calcium, sodium and aluminium in fibrous form. There are four main forms of asbestos, crocidolite (blue asbestos), amosite, chrysotile and fibrous anthophyllite, together with fibrous actinolite and fibrous tremolite.

asbestos area An area designated by an employer where the exposure to asbestos of an employee exceeds, or is liable to exceed, the control limit.
(Control of Asbestos at Work Regulations 1987)

asbestosis A fibrotic condition of the lung, resulting in scarring and thickening of lung tissue.

assignment A task or investigation undertaken to close guidelines, usually following a briefing session, in many cases to encourage transfer of learning to the actual job situation.

atomization (2C1) The process of converting a fluid to an aerosol, generally by the application of pressure on the fluid and passing same through an atomizer spray head.

attitude (1A2) This term is defined as:

- a predetermined set of responses built up as a result of experience of similar situations;
- a tendency to respond either positively or negatively to certain persons, objects or situations;

- a tendency to behave in a particular way in a particular situation.

attitude survey (2B1) A means to measure and evaluate the attitudes of certain target groups or members of the public, in order to be able to predict the future responses of such groups. For instance, the government may wish to predict public responses to impending or proposed legislation.
Companies may wish to predict responses to the introduction of new products or measure changes in attitude to existing products. An attitude survey of a selected group of people, seen to be representative of the market, may indicate this response.
Attitudes are generally measured in two ways:

- by presenting an individual with a statement and ascertaining whether or not, or to what degree, he or she accepts that statement;
- by presenting two, or a series, of options and asking an individual to select the best option.

atypical workers A term used to describe workers and others who are not in normal daytime employment, together with shift workers, part-time workers and night workers.
(Working Time Regulations 1998)

audiometry (1E3) The measurement of an individual's hearing acuity or ability over a range of frequencies.
The determination of an individual's threshold levels for pure tones by air conduction under monoaural earphone listening conditions.
An audiogram, the outcome of an audiometric test, is used to assess the degree of hearing loss across the frequencies of interest, namely the frequencies at which normal speech takes place, i.e. 0.5, 1 and 2 kHz. It is essentially a chart of a person's hearing threshold levels for pure tones of different frequencies. See Figure A.3.

auditory nerve The nerve leading from the cochlea in the inner ear and conducting impulses to the brain.

audit systems (2A4) Health and safety audit systems are designed to cover and, where appropriate, review, an organization's management systems for health and safety.
They have a number of objectives:

- to identify current strengths and weaknesses in the management structure, organization, planning and control with a view to exploiting and extending the strengths and taking remedial action to reduce or eliminate weaknesses;
- to provide a basis against which overall management effectiveness can be assessed at subsequent audits.

authoritative precedent A precedent which a judge is bound to follow. In many cases, a lower court is bound by the former decision of a higher court.

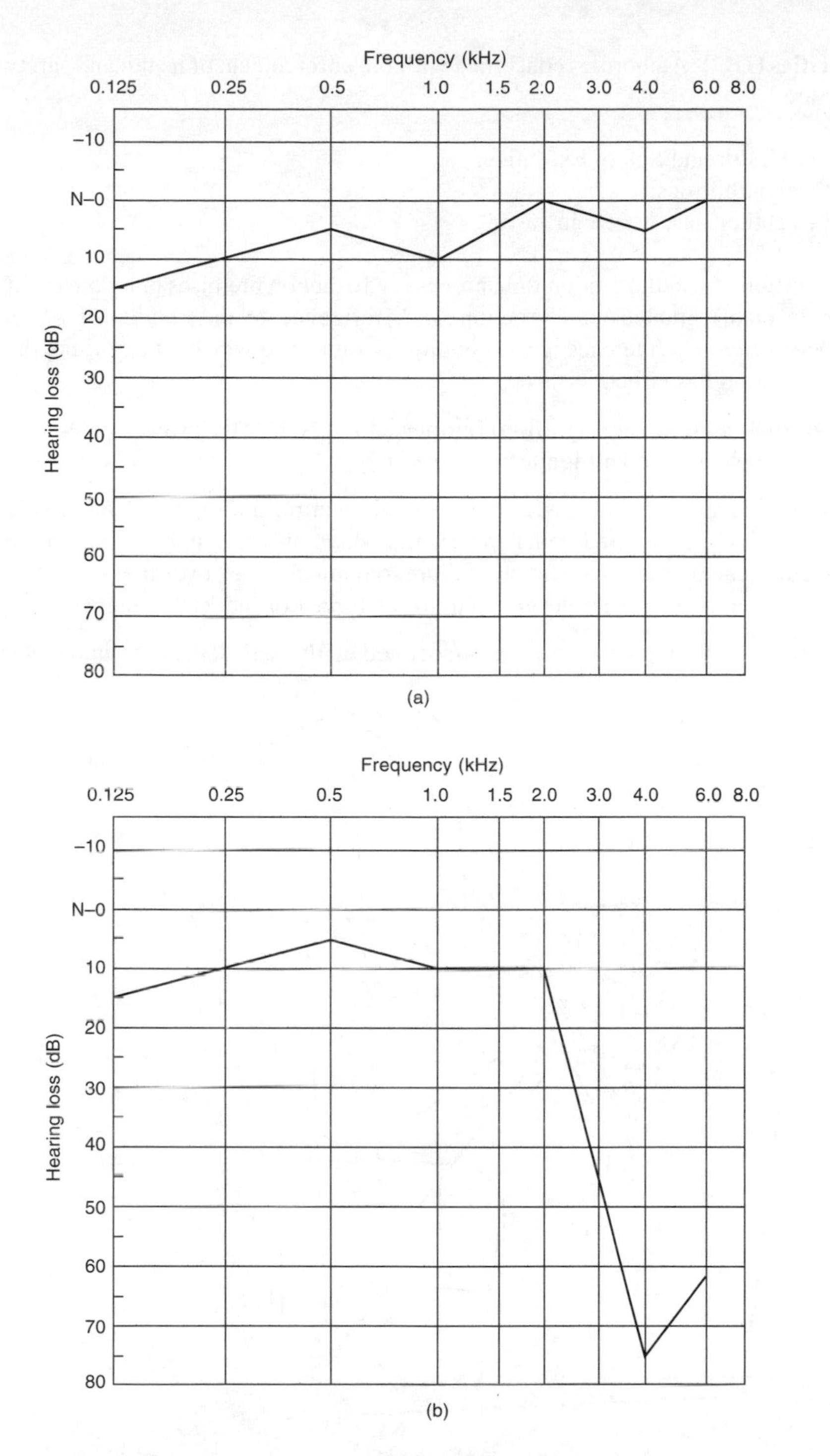

Figure A.3 Audiograms for a subject with normal hearing (a) and a subject with a moderate degree of hearing damage (b).

authorities (1B2) Authorities concerned with the enforcement of health and safety law are:

- the Health and Safety Executive;
- local authorities;
- fire authorities, in certain cases.

authorization An authorization for a process (whether on premises or by means of mobile plant) granted under section 6. A reference to the *conditions of an authorization* is a reference to the conditions subject to which, at any time, the authorization has effect.

auto-ignition (spontaneous ignition) temperature (2C1) The lowest temperature at which a substance will ignite spontaneously.

automatic firing control A device to effectively control the supply of fuel to the burners of oil or gas-fired boilers and which shuts off the supply of fuel in the event of flame or pilot flame failure, failure to ignite the fuel, over-pressurization of the boiler, loss of water level or failure of forced or induced draught fans.

automatic guard A guard which is associated with, and dependent upon, the

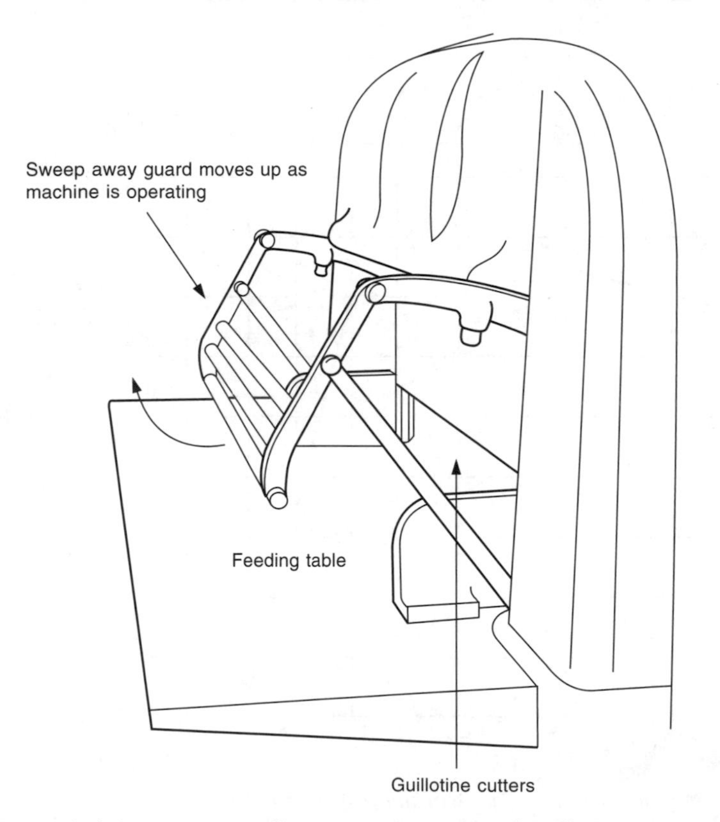

Figure A.4 Automatic guarding to a power-operated guillotine.

mechanism of the machinery and operates so as to remove physically from the danger area any part of a person exposed to danger. See Figure A.4 (BS 5304: *Safeguarding of Machinery*)

automatic water level controls A safety feature of steam boilers so arranged that they positively control the boiler or the feed pumps, or regulate the water supply to the boiler and effectively maintain the boiler water level between certain predetermined limits.

average illuminance A recommended level of illuminance in a workplace, based on the type of location or work undertaken, and listed in HS(G) 38 'Lighting at work' (HSE).

avoidance (2A3) A form of conditioned and specific emotional response whereby an individual avoids people and situations which may be stressful or contrary to their beliefs and opinions.

avoidance learning Learning to avoid a noxious stimulus, such as a shock, by responding appropriately to a warning signal.

B

balance of probabilities In a civil action, the case must be proved 'on the balance of probabilities', which is a lesser standard than that of 'beyond reasonable doubt' required in criminal cases.

base plate In tubular steel scaffolding, a flat square steel plate with a locating pin that must be inserted into the base of a vertical standard to provide a bearing surface for load distribution.

basilar membrane The membrane in the cochlea on which the organ of Corti is located. Its motion is significant in the hearing process.

'bath tub' curve (2D1) A graphical representation commonly used to demonstrate vibration as a machine condition indicator, where the length of time of operation of a machine is plotted against the vibration level of same.
The various stages between the introduction of a new machine and its eventual breakdown, through the initial running-in period, planned maintenance to the effecting of repairs, show the vibration level taking the shape of a bath tub.
Clearly, any planned maintenance programme should ensure that the right-hand side of the curve, where the vibration level is leading to breakdown, should not be allowed to rise.

beat disorders A group of prescribed occupational diseases associated with external pressure and friction at or about a particular joint. The beat disorders include:

- beat hand (subcutaneous cellulitis of the hand);
- beat knee (bursitis or subcutaneous cellulitis arising at or about the knee due to severe or prolonged external friction or pressure at or about the knee – housemaid's knee);
- beat elbow (tennis elbow).

behaviour An observable action of a person or animal.

bel (2E2) The logarithmic unit of sound pressure.

bending moment The sum of the moments along a beam of all the vertical forces acting to one side of a point about that point.

benign tumour A tumour which grows slowly and less expansively than a malignant tumour and, provided it is not located in a vital part of the body, such as the brain, is well tolerated by the body.

Best Available Techniques Not Entailing Excessive Cost (BATNEEC) Where

considering an application to operate a prescribed process, HM Inspectorate of Pollution may impose conditions with regard to the authorization. These conditions should ensure that the best available techniques not entailing excessive cost are used to prevent or, if that is not practicable, to minimize the release of prescribed substances into the environmental medium for which they are prescribed, and to render harmless any prescribed substances which might cause harm if released into the environmental medium.
(Environmental Protection Act 1990)

Best Practicable Environmental Option (BPEO) In setting the conditions within an authorization under section 7 of the Environmental Protection Act 1990, these conditions should ensure that when a process is likely to involve releases into one environmental medium, the best practicable environmental option is achieved, namely the releases from the process are controlled through the use of BATNEEC so as to have the least effect on the environment as a whole.
(Environmental Protection Act 1990)

best practicable means A person who commits an offence on industrial, trade or business premises may use the defence that he had used 'best practicable means' to prevent, or to counteract the effects of, a nuisance.
Such a defence has to be interpreted with reference to the following provisions:

(a) 'Practicable' means reasonably practicable having regard to amongst other things:
- local conditions and circumstances;
- the current state of technical knowledge; and
- the financial implications.

(b) The means to be employed include:
- the design, installation, maintenance and manner and periods of operations of plant and machinery; and
- the design, construction and maintenance of buildings and structures.

(c) The test is to apply only so far as compatible with any:
- duty imposed by law;
- with safety and safe working conditions; and
with the exigencies of any emergency or unforeseeable circumstances.

(Environmental Protection Act 1990)

beta particle A form of ionizing radiation taking the form of electrons from within the nucleus of an atom. They are ejected at great speed.

beyond reasonable doubt Where a person is charged with a criminal offence, the burden of proving the charge 'beyond reasonable doubt' rests with the prosecution. Where this cannot be done, the court must find the alleged offender 'not guilty'.

Biased Liability Theory This theory of accident causation proposes the notion that once a person has an accident, the probability that the same person will have a further accident in the future is either decreased or increased when compared with the rest of the population.

If the probability has increased, the phenomenon is known as the 'Contagion Hypothesis'. Conversely, if is has decreased, it is known as the 'Burnt Fingers Hypothesis'.

binding precedent A doctrine whereby a court must follow the earlier decisions of the courts at its own level and of superior courts. It is based on the concept of *stare decisis*, i.e. stay with the decisions of past cases.

biological agent (1E1) This means any micro-organism, cell culture or human endoparasite, including any which have been genetically modified, which may cause any infection, allergy, toxicity or otherwise create a risk to human health. (Control of Substances Hazardous to Health Regulations 1999)

Biological Exposure Index A standard set by the American Conference of Government Industrial Hygienists (ACGIH) which indicates the average body fluid concentration of a toxic substance or its metabolite found in workers exposed at the equivalent threshold limit value.

biological gradient A situation where an increase in the dose of a substance is directly related to an increase in the incidence of a disease.

biological hazards Various forms of ill health can result from exposure to biological agents, such as bacteria, viruses and dusts. Biological hazards can be classified according to origin:

- animal-borne – e.g. anthrax, brucellosis;
- human-borne – e.g. viral hepatitis;
- vegetable-borne – e.g. aspergillosis (farmer's lung).

biological indicators Absorption into the body of offending agents may be assessed by sampling and examination of certain biological indicators. The biological sampling where the indicators may be determined are of blood, urine, saliva, sweat, faeces, hair, nails and expired air.

biological matter Anything other than an entity (i.e. acellular, unicellular or multicellular entity) which consists of or includes:

- tissue or cells (including gametes or propagules) or subcellular entities, of any kind, capable of replication or of transferring genetic material; or
- genes or other genetic material, in any form, which are so capable.

It is immaterial, in determining if something is or not an organism or biological matter, whether it is the product of natural or artificial processes of reproduction and, in the case of biological matter, whether it has ever been part of a whole organism.
(Environmental Protection Act 1990)

biological monitoring A regular measuring activity where selected validated indicators of the uptake of toxic substances are determined in order to prevent health impairment.

Biological monitoring may feature as part of the health surveillance procedures required under the Control of Substances Hazardous to Health (COSHH) Regulations 1999.

It may be undertaken through the determination of the effects certain substances produce on biological samples of exposed individuals, and these determinations are used as biological indicators.

Biological samples where indicators may be determined consist of:

- blood, urine, saliva, sweat, faeces;
- hair, nails; and
- expired air.

Indicators of internal dose can be divided into:

- true indicators of dose, i.e. capable of indicating the quantity of the substance at the sites of the body where it exerts its effect;
- indicators of exposure, which can provide an indirect estimate of the degree of exposure, since the levels of substances in the biological samples closely correlate with levels of environmental pollution; and
- indicators of accumulation that can provide an evaluation of the concentration of the substance in organs and/or tissues from which the substance, once deposited, is slowly released.

biological plausibility This term is used to describe the relationship between the cause and effect of a disease, which should not conflict with the established facts as to the aetiology of the disease.

Biological Tolerance Value This is a standard established in Germany and defined as 'the maximum permissible quantity of a chemical compound, or its metabolites, or any deviation from the norm of biological parameters induced by these substances in exposed humans'.

Such levels do not impair human health, even if the exposure is repeated and of long duration, and are, in effect, health-based values or acceptable upper limits.

blow down valve A valve, commonly located near the surface of the water in a boiler shell, which has the following functions:

- deconcentration of the boiler water to prevent the solids content rising above prescribed limits;
- the ejection of solids and sludge precipitated from boiler water which tend to settle along the bottom of a boiler;
- emptying the boiler prior to inspection or for other purposes.

An intermittent blow down valve is commonly fitted at the lowest part of a boiler. It is commonly used for the systematic ejection of unwanted solid matter deposited from the boiler water.

body protection (1E4) This form of personal protective equipment includes one-

piece and two-piece overalls, aprons, donkey jackets, foul weather clothing and overtrousers.
The selection of appropriate body protection is required under the Personal Protective Equipment at Work Regulations 1992. The following factors should be considered:

- the ease and cost of washing or dry cleaning;
- the degree of exposure to personal contamination from work activities, wet processes and variations in humidity and temperature;
- possible discomfort during hot weather;
- ease of storage.

Boyle's law The volume of a given mass of gas is directly proportional to its absolute temperature as long as the pressure remains constant.

brace In tubular steel scaffolding, a tube fixed diagonally across a scaffold or scaffold bay to prevent lateral movement and improve stability of the scaffold.

brainstorming Sometimes referred to as 'thinking out loud', this is a highly intensive technique for gathering ideas, views and suggestions from a pre-selected group on a particular issue, and in problem-solving activities. The technique endeavours to rid participants of any inhibitions to thinking.

brainwashing Systematic attempts to change attitudes.

brake A device which is interlocked with a guard and machine controls so that the act of cutting off the power to the dangerous parts or opening the guard applies the brake. A brake is virtually instantaneous in action.

breach of statutory duty (1B1) In certain cases a breach of a criminal duty imposed by a statute or regulations may give rise to civil liability and a resultant claim for damages.
In many cases, a statute or regulations will state that a breach of that statute or regulations shall not give rise to civil liability for injury, damage or loss (e.g. Management of Health and Safety at Work Regulations 1999), but where such a statement is not made civil liability is implied.
The standard test adopted by the courts has been to ask the question:

> 'Was the duty imposed specifically for the protection of a particular class of person, or was it intended to benefit the public at large?'

If the answer to the first part of the question is 'yes', a civil claim may be allowed. The courts have always viewed legislation, such as the safety provisions of the Factories Act 1961, as being directed towards protecting a particular class, and have allowed civil claims for damages by persons belonging to these protected classes, namely employees.

breakdown maintenance (2A3) Maintenance undertaken following the breakdown of an item of equipment to restore it to an acceptable condition.

breaking stress (2D4) The stress in a material at which fracture takes place.

breathing apparatus Apparatus which replaces the surrounding air supply and provides the user with sufficient air to breathe normally.
Air line breathing apparatus consists of a full face mask or half-mask connected by a flexible hose either to a source of uncontaminated air (short distance) or to a compressed air line via a filter and demand valve.
Self-contained breathing apparatus can be:

- open circuit, which supplies air by a lung-governed demand valve or pressure reducer connected to a full face piece via a hose supply, the hose being connected to its own compressed air or oxygen supply carried by the user in a harness;
- closed circuit, which incorporates a purifier to absorb exhaled carbon dioxide, the purified air being fed back to the respirator after mixing with pure oxygen.

Bremsstrahlung A form of ionizing radiation which are weak X-rays produced by negative beta particles impinging on heavy materials.

brightness (luminosity) (2E3) The perceived intensity of light, which is very much a subjective sensation and cannot be measured. It is possible, however, to use a Brightness Ratio.

Brightness Ratio This is the ratio of apparent luminosity between a task object and its surroundings.
To achieve a recommended Brightness Ratio, the reflectance of all surfaces in the workplace should be carefully maintained and consideration given to reflectance values in the design of interiors.
Given a task illuminance factor of 1, the following effective reflectance values are recommended:

Ceilings	0.6
Walls	0.3–0.8
Floors	0.2–0.3.

British Standards The British Standards Institution produces safety standards and codes through committees formed to deal with a specific matter or subject, such as machinery safety.
Standards contain details relating to, for instance, the construction of and materials incorporated in an item and, where necessary, prescribe methods of testing to establish compliance.
Codes deal with safe working practices and systems of work.
British Standards and Codes have no legal status, but can be interpreted by the courts as being the authoritative guidance on a particular matter.

brittle fracture (2D5) This form of failure is commonly the cause of pressure vessel explosions. It is associated with:

- the use of inherently brittle metals, such as cast iron, or the embrittlement of

other metals arising from exposure to excessive heat, particularly in the case of welding;
- the effects of very low temperatures, which can bring about a change in the structure of materials;
- impact or shock loading, particularly in the case of lifting appliances;
- built-in stresses during fabrication of metal structures;
- the presence of notches and material defects.

burden (onus) of proof (1B1) Criminal law generally requires the prosecution to prove beyond a reasonable doubt that a person committed an offence.
Under the HSWA, the burden of proof rests with the accused to prove that it was not practicable or reasonably practicable or there were no better practicable means to satisfy the duty or requirement. In a trial for a breach of the HSWA and/or regulations, the prosecution need only prove that a breach of the Act and/or regulations took place, and that the accused was responsible for same.
Where the accused cannot prove that it was not practicable or reasonably practicable or that there were no better practicable means, the case would be considered proved against him or her.

Burnt Fingers Hypothesis See ***Biased Liability Theory***.

Business Interruption An important element of a Total Loss Control programme, this entails the incorporation into the programme of controls over all situations and influences which downgrade the system and result in business interruption, e.g. product liability procedures, pollution prevention.

C

canister respirator A form of gas respirator, incorporating goggles and visor, with an exhalation valve and connected to a chemical canister filter. This respirator is suitable for protection against low concentrations of designated toxic gases and vapours.

cantilever A beam supported at one end only.

capacity In order to be able to enter into a contract, both parties must have legal capacity. This means they must be over the age of 18, sane and sober.

captor system A form of local exhaust ventilation (LEV) system, whereby the moving air captures the contaminating agent at some point outside the collection hood and induces its flow into the hood.

carbon dioxide appliance A form of fire appliance consisting of a pressure cylinder filled with liquid carbon dioxide. When the trigger is pressed, carbon dioxide 'snow' is emitted through the horn, which rapidly converts to carbon dioxide gas in a fire.
Carbon dioxide can be used on both Class A and Class B fires and on fires involving electrical equipment.

carcinogen (2E1) This means:

(a) any substance or preparation which, if classified in accordance with the classification provided for by regulation 5 of the Chemicals (Hazard Information and Packaging for Supply) Regulations 1994, would be in the category of danger, carcinogenic (Category 1) or carcinogenic (Category 2) whether or not the substance or preparation would be required to be classified under those Regulations; or
(b) any substance or preparation:
(i) listed in Schedule 1; and
(ii) any substance or preparation arising from a process specified in Schedule 1 which is a substance hazardous to health.

(Control of Substances Hazardous to Health Regulations 1999)

carcinogenesis (2E1) The process of uptake in the body of carcinogenic substances.

carcinogenic (classification) Substances and preparations which, if they are inhaled or ingested or if they penetrate the skin, may induce cancer or increase its incidence.

[Chemicals (Hazard Information and Packaging for Supply) Regulations 1994]

cartridge respirator A respirator which incorporates a chemical cartridge filter and which may be effective against low concentrations of relatively non-toxic gases or vapours which have an acceptable level of concentration exceeding 100 ppm.

case control study (2E5) This type of study is undertaken in order to compare the frequency of a particular occupational disease or condition against a control group drawn from the population at large. This may also be used as a preliminary study, to ascertain whether there could be a link between a suspected cause and known effect, and in the investigation of diseases which are not particularly prevalent.

case law The written decisions of the criminal and civil courts.
This is based on the doctrine of judicial precedent, and its principles and doctrines are to be found in the law reports, e.g. All England Reports. It is a self-endorsing process, perpetuated either by previous binding cases or by the interpretation of legislation.
The following features of a judgement are important:

- The *ratio decidendi* (reason for deciding) – a statement of law based on an examination of the facts and the legal issues surrounding them; this is the most important part of a judgement and contains the actual binding precedent.
- The *obiter dicta* (words said by the way) – may contain a statement about the law which is not based on the facts of the case under review and which will not, therefore, be part of the decision; this is often held to be of persuasive authority.

case study A problematic situation presented to a trainee or group of trainees with a view to that trainee or group finding the best solution, usually in the form of a written report or oral presentation following the study.

category of danger Under the Chemicals (Hazard Information and Packaging for Supply) Regulations 1994, substances and preparations may need to be marked with a category of danger, details of which are indicated in Column 1 of Schedule 1 to the regulations.
Categories of danger are:

- Physico-chemical – Explosive, Oxidizing, Extremely Flammable, Highly Flammable and Flammable.
- Health effects – Very Toxic, Toxic, Harmful, Corrosive, Irritant, Sensitizing (three categories), Carcinogenic (three categories), Mutagenic (three categories), Toxic for Reproduction.
- Environment – Dangerous for the Environment (substances only).

Cause–Accident–Result Sequence A theory of accident causation which proposes that the indirect causes of an accident (personal factors and source causes) contribute to the direct causes (unsafe acts and unsafe conditions) which result

INDIRECT CAUSES

PERSONAL FACTOR
Definition: Any condition or characteristic of a man that causes or influences him to act unsafely.

(1) Knowledge and skill deficiencies
- (a) Lack of hazard awareness
- (b) Lack of job knowledge
- (c) Lack of job skill

(2) Conflicting motivations
- (a) Saving time and effort
- (b) Avoiding discomfort
- (c) Attracting attention
- (d) Asserting independence
- (e) Seeking group approval
- (f) Expressing resentment

(3) Physical and mental incapacities

SOURCE CAUSES
Definition: Any circumstances that may cause or contribute to the development of an unsafe condition.

MAJOR SOURCES
(1) Production employees
(2) Maintenance employees
(3) Design and Engineering
(4) Purchasing practices
(5) Normal wear through use
(6) Abnormal wear and tear
(7) Lack of preventive maintenance
(8) Outside contractors

DIRECT CAUSES

UNSAFE ACT
Definition: Any act that deviates from a generally recognized safe way of doing a job and increases the likelihood of an accident.

BASIC TYPES
(1) Operating without authority
(2) Failure to make secure
(3) Operating at unsafe speed
(4) Failure to warn or signal
(5) Nullifying safety devices
(6) Using defective equipment
(7) Using equipment unsafely
(8) Taking unsafe position
(9) Repairing or servicing moving or energized equipment
(10) Riding hazardous equipment
(11) Horseplay
(12) Failure to use protection

UNSAFE CONDITIONS
Definition: Any environmental condition that may cause or contribute to an accident.

BASIC TYPES
(1) Inadequate guards and safety devices
(2) Inadequate warning systems
(3) Fire and explosion hazards
(4) Unexpected movement hazards
(5) Poor housekeeping
(6) Protruding hazards
(7) Congestion, close clearance
(8) Hazardous atmospheric conditions
(9) Hazardous placement or storage
(10) Unsafe equipment defects
(11) Inadequate illumination, noise
(12) Hazardous personal attire

THE ACCIDENT
Definition: An unexpected occurrence that interrupts work and usually takes this form of an abrupt contact.

BASIC TYPES
(1) Struck by
(2) Contact by
(3) Struck against
(4) Contact with
(5) Caught in
(6) Caught on
(7) Caught between
(8) Fall to different level
(9) Fall on same
(10) Exposure
(11) Overexertion/strain

DIRECT RESULTS
Definition: The immediate results of an accident.

BASIC TYPES
(1) 'No results' or near miss
(2) Minor injury
(3) Major injury
(4) Property damage

INDIRECT RESULTS
Definition: The consequences for all concerned that flow from the direct result of accidents.

FOR THE INJURED
(1) Loss of earnings
(2) Disrupted family life
(3) Disrupted personal life
(4) And other consequences

FOR THE COMPANY
(1) Injury costs
(2) Production loss costs
(3) Property damage costs
(4) Lowered employee morale
(5) Poor reputation
(6) Poor customer relations
(7) Lost supervisor time
(8) Product damage costs

Figure C.1 The cause–accident–result sequence.

in the accident. The accident has direct results, that is the immediate results of the accident (such as personal injury and property damage), but also indirect results, both for the injured person (such as loss of earnings) and for the organization (such as poor reputation). See Figure C.1.

caustic cracking A form of stress corrosion which takes the form of a stress-produced crack in boilers. It is associated with local concentrations of sodium hydroxide present in boiler feedwater which attack steel at high temperatures.

cavity barrier A barrier capable of restricting the movement of flame and smoke within a structural element of a building.

CE marking (2D2) A specific form of marking which must be applied to a wide range of equipment, such as electrical equipment, indicating that the equipment complies with all the requirements of regulations, e.g. the Electrical Equipment (Safety) Regulations 1994, which implement that particular European Council Directive.

Change Analysis Part of the Management Oversight and Risk Tree (MORT) system, this is based on the Kepner–Tregoe method of rational decision making. Change analysis compares a problem-free situation with a problem (accident) situation in order to isolate causes and effects of change.
The technique is particularly effective when the decision maker needs a quick analysis, when the cause is obscure, and when well-behaved operators behave differently from previous situations, e.g. the Three Mile Island incident.

change of process A common strategy in protecting both the health and safety of people exposed to hazards. In this case, improved design or process engineering can result in changes to provide better protection, as in the case of dusty processes or those producing noise.

Charles' law The volume of a given mass of gas is directly proportional to its absolute temperature as long as the pressure remains constant.

chemical asphyxiant A substance which exerts a physiological effect on the blood and body tissue by combining with haemoglobin, thereby preventing or greatly reducing its function as a carrier of oxygen.

chemical foam extinguisher An appliance incorporating a cylinder containing two solutions, aluminium sulphate and sodium bicarbonate, which are mixed on inversion of the cylinder. The foaming mixture is expelled in the inverted position by carbon dioxide generated in the chemical reaction.

chemical hazards These are hazards arising from the use and storage of chemical substances, and which result in a range of chemical poisonings and other forms of disease or condition. Exposure to these hazards may result in dermatitis, occupational cancers and respiratory disorders.

chemical reaction (2C3) Such a reaction takes place when the atoms in molecules are rearranged, either as a result of breaking down (decomposition) into smaller molecules, or by the combining of different molecules to form different structures.

Chemical reactions may be exothermic, whereby heat is emitted, or endothermic, where heat is absorbed from the surroundings.

child A person under compulsory school leaving age and under 16 years.

chronic effect An effect on the body as a result of prolonged exposure or repeated exposure of long duration.

ciliary escalator An important protective mechanism in the human respiratory system, whereby cilia, which line the trachea, escalate foreign bodies and dust particles, which become lodged in the cilia, back to the mouth, where they can be expectorated or swallowed.

circuit The means for the passing of electric current, comprising a source of current, conductors to convey the current and the appliance or item of equipment powered by electrical current. The current returns to the source via the neutral or return conductor.

circuit breaker A device incorporating a mechanism that trips a switch from the ON to OFF position where an excess of current may flow in a circuit. It should be of the type and rating for the circuit and appliance it is designed to protect.

circuit conductor Any conductor in a system which is intended to carry electric current in normal conditions, or to be energized in normal conditions, and includes a combined neutral and earth conductor, but does not include a conductor provided solely to perform a protective function by connection to earth or other reference point.
(Electricity at Work Regulations 1989)

circulatory system (1E1) An important body system comprising the heart, arteries and veins. The heart provides a pumping action, pumping blood, which contains oxygen received from the lungs, via the arteries, to all parts of the body. Carbon dioxide is removed from tissue and carried back in blood through veins to the lungs, where it is eventually exhaled.
The heart is, fundamentally, divided into four chambers, a left and right auricle and a left and right ventricle. These chambers are connected by valves which regulate the flow of blood through the heart. The left ventricle is stronger and thicker than the right ventricle due to its principal function of pumping blood to all parts of the body. The heart beats at around 70 beats per minute, but this may be increased during exercise or slowed during sleep or rest.

civil action (1B1) In such legal actions a claimant sues a defendant, commonly within the tort of negligence, and pursues a claim for damages in respect of personal injury, damage, loss or death.

civil court procedure (1B1) Most civil cases are heard by a judge sitting alone. The first stage in civil proceedings is the issue of a writ of summons or writ, which must be used for any action in *tort* (other than trespass on to land) and for any claim for damages in respect of death, personal injury or damage to property. The defendant acknowledges service of writ within 14 days of service to either the Central Office of the High Court or the appropriate District Registry.

The claimant counsel outlines allegations against the defendant, the particulars of which have been given to the defendant by way of a statement of claim.
Witnesses are called and give evidence for both sides, where appropriate.
Defence counsel can submit:

- that there is no case to answer;
- that he or she intends to call no evidence; or
- his or her defence with evidence.

The judge finds for or against the claimant, awards the appropriate remedy (e.g. damages) and levies costs.
Under the Courts and Legal Services Act 1990, actions in the County Court are designed for cases where damages not exceeding £50 000 are involved.
A summons is requested by the claimant, which sets out the details of the parties to the action and the nature and amount of the claim.
The registrar prepares the summons and serves it on the defendant or his solicitor, who must accept it.
The defendant may:

- pay the whole sum claimed plus costs;
- ask for time to pay;
- file a defence if he or she denies liability;
- dispute the claim on the return day, but may have to pay extra costs if he or she does so.

civil liability (1B1) Civil liability refers to the 'penalty' that can be imposed by a civil court.
A civil action generally involves individuals, with a claimant suing a defendant for a remedy or remedies; in most cases the remedy takes the form of damages, a form of financial compensation.
A civil case must be proved on the 'balance of probabilities', which is a lesser standard than that of 'beyond reasonable doubt', which is required in a criminal case.
Many claims for injuries sustained at work pass through the civil courts or may be settled out of court.
The civil courts are the County Court and the High Court.

Class 1 appliances Portable electrical appliances which incorporate both earthing and insulation (earthed appliances).

Class 2 appliances Portable electrical appliances which are doubly insulated.

classical conditioning Learning that takes place when a conditioned stimulus is paired with an unconditioned stimulus.

Classical Organization Theory This theory by Bass (1965) is based on Theory X. See individual entry. It can be summarized in eight precepts:

1. Some one person should be responsible for each activity.
2. Responsibility for each activity should not be duplicated and should not overlap.
3. Each position should have a limited number of clearly stated duties.
4. Every person should know exactly what their duties are.
5. Authority for making decisions should be commensurate with the responsibility for them.
6. Authority should be delegated to persons close to the point of action.
7. Managers should have a limited number of subordinates, e.g. four to seven.
8. Each manager should know to whom he or she reports and who reports to him or her; the chain of command should be clearly defined.

cleaning schedule Poor or inadequate levels of cleaning and housekeeping are a contributory factor in many accidents.
Management of the cleaning operation is best undertaken by the operation of a formal cleaning schedule which incorporates the following information:

- the item of plant or equipment or area to be cleaned;
- the cleaning procedure to be followed;
- the frequency of cleaning, e.g. daily, weekly, at the end of each production run;
- individual responsibility for ensuring the cleaning standard is achieved;
- any precautions necessary, e.g. in the use of hazardous cleaning preparations.

cleanliness of the workplace Employers have an absolute duty to ensure that every workplace, and the furniture, furnishings and fittings therein, are kept sufficiently clean.
Surfaces of floors, walls and ceilings of all workplaces inside buildings must be able to be kept sufficiently clean.
So far as is reasonably practicable, waste materials must not be allowed to accumulate except where stored in suitable receptacles.
[Workplace (Health, Safety and Welfare) Regulations 1992]

client A person for whom a project is carried out, whether it is carried out by another person or carried out in-house.
[Construction (Design and Management) Regulations 1994]

cochlea A bony cavity, coiled like the shell of a snail, containing receptor organs for hearing.

cognitive dissonance A motivational state produced by inconsistencies between simultaneously held cognitions (thoughts or ideas) or between a cognition and behaviour.
The conflict situation that can arise where a person holds an attitude which is not compatible with the information presented.
Festinger expressed the notion that people, when presented with two pieces of

information [e.g knowledge, thoughts, feelings (cognitions)] that are inconsistent, opposite or conflicting, express discomfort. In order to reduce this discomfort, or dissonance, they engage in a variety of activities by avoiding such thoughts or by changing their ideas, knowledge or skills.

cognitive overload A state in which there is more information directed at a person than they can process in thought at a particular time.

Cohort Studies (2E5) This is a form of epidemiological investigation of a retrospective nature which is used to identify whether there is a link between a particular exposure to, for instance, a hazardous substance (the cause) and an occupational disease (the effect). Two groups or cohorts are used in this study, namely a group of people who have not been exposed (the control group) and those who have been exposed to the substance (the exposed group).
The incidence of the disease being monitored is calculated for both groups. Where significantly more people are identified as suffering from the disease in the exposed group, there may well be strong evidence to show a cause and effect. Generally, this form of study can provide an accurate assessment of exposure related to death or disease. On this basis, a direct estimate of the risk related to causal factors can be made.

collision energy The physical representation of the initiation energy required for a fire to start.

colour rendition The appearance of an object under a given light source compared to its colour under a reference illuminant, such as natural light. Colour rendition enables a colour to be correctly perceived.

combustion The process of burning; the combustion process can be written as:

$$\text{Fuel} + \text{Oxygen} = \text{Products of combustion} + \text{Heat}$$

Combustion is, fundamentally, an oxidation reaction whereby the carbon element of fuel is oxidized to carbon dioxide and the hydrogen element to water.

comfort A subjective assessment of the conditions in which a person works, sleeps, relaxes, travels, etc. and which varies according to age, state of health and vitality. Comfort is directly related to environmental factors such as temperature, ventilation and humidity.

comfort ventilation The process of providing sufficient air for people to breathe and, to some extent, regulating temperature. It is directly related to the number of air changes per hour in a workplace according to the external ambient air temperature and the actual rate of air movement. Rates of air change will, in most cases, vary from summer to winter in order to maintain comfort.

common law (1B1) Common law is the body of accumulated case law which is:

- based on the decision of the courts over many years whereby precedents are established;
- of universal application;

- recorded in the various Law Reports, e.g. the All England Reports (AER);
- applicable to the decisions made by courts at their own level and from superior courts.

communication The transfer of information, ideas, feelings, knowledge and emotions between an individual, or group of individuals, and another. The principal function of communication is to convey meaning.

communication cycle (1CS1) An essential part of the communication process which takes place in six stages:

1. Conceiving the message.
2. Encoding.
3. Selecting the medium for transmission.
4. Decoding of the message by the recipient.
5. Interpreting the message.
6. Feedback, which confirms that the transmitter of the message has correctly received the information transmitted.

communication structure The pattern of open and closed channels of communication within a group of people.

compartmentation A structural process, designed to limit the spread of fire within a building, which divides the building into fire-resistant cells or units, both vertically and horizontally.
It is further used to segregate high risk areas of a building from other areas.

compensation A defence mechanism in which an individual substitutes one activity for another in an attempt to satisfy frustrated motives. Compensation commonly implies failure or loss of self-esteem in one activity and the compensation of this loss by efforts in some other aspect of endeavour.

competent person (2C2) The term 'competent person' is used in a wide range of legislation, e.g. Management of Health and Safety at Work Regulations 1999, Construction (Health, Safety and Welfare) Regulations 1996, but is not generally defined in statutes and regulations.
The term was defined in Brazier v. Skipton Rock Co. Ltd. (1962) as meaning a person with practical and theoretical knowledge, as well as sufficient experience of the particular machinery, plant or procedure involved, to enable him or her to identify defects or weaknesses during plant and machinery examinations, and to assess their importance in relation to the strength and function of that plant and machinery.
Examples where employers must appoint competent persons are:

- to assist them in undertaking the measures they need to take to comply with the relevant statutory provisions;
- to implement procedures for serious and imminent danger and for danger areas (Management of Health and Safety at Work Regulations 1999);

- examination and testing of chains, ropes, lifting tackle, lifting machines and lifts (Lifting Operations and Lifting Equipment Regulations 1998);
- making noise assessments (Noise at Work Regulations 1989);
- preparation of Written Scheme of Examination for the examination of pressure systems (Pressure Systems Regulations 1999);
- supervision of various construction operations, such as the erection of a scaffold, demolition of a structure [Construction (Health, Safety and Welfare) Regulations 1996].

compliance Behaviour in accordance with group pressures without necessarily accepting the values and norms of the group.

compound stresses Stresses arising in materials, other than tension and compression and direct shear, the most common being torsional stress, which is produced by twisting, and transverse stress, which arises as a result of bending.

compressive stress A form of stress applied to structures and/or materials where the forces applied act along the same line but towards each other, producing a 'squashing' effect.

$$\text{Compressive stress} = \frac{\text{Force applied}}{\text{Cross-sectional area of the material}}$$

conditioned response A response produced by a conditioned stimulus after learning.

condition monitoring A system for evaluating the physical state of machinery and equipment over a period of time using a range of techniques, such as strain measurement, corrosion monitoring, vibration measurement and acoustic emission monitoring.

conduction The method of heat transfer in solids.

conductivity (thermal or electrical) The measure of the ability of a substance to allow the passage of electricity or heat.

conductor A conductor of electrical energy.
(Electricity at Work Regulations 1989)
A solid capable of transferring heat.

confined space A place which is substantially, though not always entirely, enclosed, and where there is a risk that anyone who may enter the space could be injured due to fire or explosion, overcome by gas, fumes, vapour or the lack of oxygen, drowned, buried under free-flowing solids, such as grain, or overcome due to high temperature.
Under the Confined Spaces Regulations 1997, employers must:

(a) prevent employees from entering confined spaces – for example, by undertaking the work from outside;
(b) follow a safe system of work (e.g. a Permit to Work system) if entry to a confined space is unavoidable; and

(c) put in place adequate emergency arrangements before work starts, which will also safeguard any rescuers.

The regulations are accompanied by an ACOP and HSE Guidance.

conflict (1CS1) This implies a clash, struggle, a trial of strength involving two or more persons or groups.
Sources of conflict within an organization, in particular between the line organization and the staff organization, include:

- differing motivations;
- misunderstanding of individual roles;
- differing cultures and objectives;
- differing priorities and levels of commitment.

conformity The tendency to be influenced by group pressure and to acquiesce to group norms.

Consequence Analysis A feature of risk analysis which considers the particular effects of a particular process failure and the damage caused by these effects. Consequence analysis is undertaken to form an opinion on potentially serious hazardous outcomes of accidents and their possible consequences for people and the environment. It should act as a tool in the decision-making process in a safety study.

consideration That legal element that changes an informal agreement between parties into a legally binding contract, e.g. goods for money, work for money.

construction work The carrying out of any building, civil engineering or engineering construction work, including any of the following:

(a) the construction, alteration, conversion, fitting out, commissioning renovation, repair, upkeep, redecoration or other maintenance (including cleaning), which involves the use of water or an abrasive at high pressure or the use of substances classified as corrosive or toxic for the purposes of Regulation 7 of the Chemicals (Hazard Information and Packaging for Supply) Regulations 1994, decommissioning, demolition or dismantling of a structure;
(b) the preparation for an intended structure, including site clearance, exploration, investigation (but not site survey) and excavation, and laying or installing the foundations of a structure;
(c) the assembly of prefabricated elements to form a structure or the disassembly of prefabricated elements which, immediately before such disassembly, formed a structure;
(d) the removal of a structure or part of a structure or of any product or waste resulting from demolition or dismantling of a structure or from disassembly of prefabricated elements which, immediately before such disassembly, formed a structure;
(e) the installation, commissioning, maintenance, repair or removal of mechanical,

electrical, gas, compressed air, hydraulic, telecommunications, computer or similar services which are normally fixed within or to a structure.

[Construction (Design and Management) Regulations 1994]

contact hazard A hazard arising from contact with a machine at a particular point arising from sharp surfaces, sharp projections, heat and extreme cold.

contagion hypothesis See ***Biased Liability Theory***.

containment/enclosure (2C3) The structural prevention of fire spread to other parts of a premises through the use of fire doors, fireproof compartments and other fire-resistant structures.
Total containment or enclosure of a dangerous processing operation is effected by the use of bunds, bulk tanks and pipework to deliver a liquid directly into a closed production vessel.

containment testing A technique for assessing the effectiveness of exhaust ventilated enclosures, whereby a tracer aerosol, such as a gas, is released at a predetermined rate inside the enclosure and then monitored at set points outside the enclosure to identify escapes from same.

contingency plan/emergency procedure (2A3) The Management of Health and Safety at Work Regulations 1999 require employers to:

(a) establish and where necessary give effect to appropriate procedures to be followed in the event of serious and imminent danger to persons at work in their undertaking;
(b) nominate a sufficient number of competent persons to implement those procedures in so far as they relate to the evacuation from the premises of persons at work in their undertaking; and
(c) ensure that none of their employees has access to any area occupied by them to which it is necessary to restrict access on the grounds of health and safety unless the employee concerned has received adequate health and safety instruction.

The extent and detail of a contingency plan or emergency procedure will depend upon the risks arising in the workplace, together with any risks to surrounding communities.
A written contingency plan should take account of the four principal phases of an emergency situation:

- Phase 1 – Preliminary Action.
  This should take into account the preparation of a plan including a list of all key telephone numbers, the system for providing emergency lighting, designation of exit routes, a plan indicating hydrant points and the location of shut-off valves to energy supplies, familiarization of all employees with details of the plan, together with the initiation of a programme of inspection of potentially hazardous areas, testing of warning systems and evacuation procedures.

- Phase 2 – Action when Emergency is Imminent.
  Where a warning of an imminent emergency is possible, this period should be used to assemble key personnel, to review standing arrangements, to give advance warning to external authorities, and test all systems connected with the emergency plan.
- Phase 3 – Action during the Emergency.
  If Phase 1 has been correctly implemented, and Phase 2 where appropriate, Phase 3 should proceed according to plan. However, in the event of unexpected variations in the predicted emergency, decision-making personnel, selected beforehand for this purpose, must be available to make precise and rapid judgements and decisions.
- Phase 4 – Ending the Emergency.
  There must be a procedure for declaring plant, systems and specific areas safe, together with an early re-occupation of buildings where possible.

Other factors to be considered in the development of a contingency plan are:

- liaison with external authorities;
- the appointment of an emergency controller and establishment of an emergency control centre;
- specification of individual authority for starting the procedure;
- notification to local authorities;
- call-out of competent persons;
- immediate action on site;
- evacuation procedure;
- access to records;
- external communication arrangements;
- public relations procedures;
- catering and temporary shelter arrangements;
- contingency repair and reinstatement arrangements;
- training of key personnel, competent persons and other employees.

contract law (2B4) A contract is a legally binding agreement between two parties. It consists of an offer made by one party that must be accepted unconditionally by the other. There must be 'consideration' that passes from the first party to the second party, that is the legal component that changes an informal agreement into a legally binding contract, such as goods for money, work for wages or accommodation for rent.
There must be 'intention', whereby the parties concerned must intend to enter into this legally binding agreement.
Parties must have legal 'capacity', i.e. be sane, sober and over the age of 18 years.

contract of employment (2B4) A contract between an employer and employee, which follows:

- an offer of a job which is accepted;

- terms of employment, which are expressed (stated verbally and/or in writing), implied (by custom and practice) and may be incorporated (through collective agreements).

Under the Employment Protection (Consolidation) Act 1978, the contract must incorporate 'a written statement of particulars' specifying parties to the contract, date of employment, rate of payment, hours of work, benefits, job title, length of notice and other conditions.

contractor A person engaged to perform a certain task without direction from the person employing him or her. This implies a certain degree of independence from the person employing him or her. The basic test of whether a person is an independent contractor is one of control over the carrying out of the work detailed in the contact.

contributory negligence (1B1) Where a person suffers damage as the result partly of his or her own fault and partly of the fault of any other person or persons, a claim in respect of that damage shall not be defeated by reason of the fault of the person suffering the damage, but the damages recoverable shall be reduced to such extent as the court thinks just and equitable having regard to the claimant's share in the responsibility for the damage.

control interlocking A form of electrical interlocking for machinery guards which incorporates an actuating switch operated by the guard, interposed electromechanical relays and/or solid-state switching devices, an electromagnetic contactor (or solid-state equivalent, e.g. thyristor) and/or a pneumatic or hydraulic solenoid valve controlling the power to the drive.
Failure of any of these elements, or of the wiring connecting them, can all be failures to danger.

controlled operation A form of protection and control strategy which may include isolation of processes in which hazardous substances are used, the use of mechanical or remote-controlled handling systems for radioactive substances, the use of Permits to Work and the restriction of certain activities to highly trained operators.

control limits (asbestos) Concentrations of asbestos in the atmosphere as follows:

- 0.2 fibres per millilitre of air averaged over any continuous period of 4 hours (crocidolite and amosite);
- 0.6 fibres per millilitre of air averaged over any continuous period of 10 minutes (crocidolite and amosite);
- 0.5 fibres per millilitre of air averaged over any continuous period of 4 hours (asbestos other than crocidolite and amosite);
- 1.5 fibres per millilitre of air averaged over any continuous period of 10 minutes (asbestos other than crocidolite and amosite).

(Control of Asbestos at Work Regulations 1987)

convection The means of heat transfer in liquids and gases by the circulation of convection currents.

cooling The most common means of fighting a fire using water. This results in a loss of reignition energy and vaporization of the water to steam, which has a blanketing or inerting effect, slowing down or preventing reignition of fuel.

corporate liability Corporate liability refers to the liability of all those directing an undertaking, namely the corporate body.
A corporate body (e.g. the directors, chief executive, etc. of a company) may be liable for most criminal offences providing a fine is specified for the offence, the offence is committed by a 'controlling mind', such as a managing director, company secretary or director, and it is committed in the course of his or her corporate duties.
Section 37 of the HSWA makes provision for a constituted board of directors to be prosecuted where an offence is committed through neglect or omission on their part. Where a functional director is guilty of an offence, he or she can personally be prosecuted in addition to the organization.

corrective maintenance Maintenance carried out to restore (including adjustment and repair) an item which has ceased to meet an acceptable condition.

corrosion (2D4) Corrosion of metals takes place as a direct result of chemical attack and through oxidation at high temperatures. This may be associated with atmospheric pollution and contact with chemical substances in a processing operation, such as acids and alkalis.
Chemical attack results in a chemical conversion process, the metal being converted into a compound, resulting in a weakened structure.

corrosive (1E1) Substances and preparations which may, on contact with living tissue, destroy it.
[Chemicals (Hazard Information and Packaging for Supply) Regulations 1994]

cosmic rays These are high-energy ionizing radiations from outer space, having a complex composition at the surface of the Earth.

cost-benefit analysis (2A1) The process of quantifying the costs of a particular strategy in, for instance, accident prevention, against the benefits to be derived through reduced accidents, lost time and property damage.

Council for Registered Gas Installers (CORGI) This organization was established in 1991. Under the Gas Safety (Installation and Use) Regulations 1994, any employer of, or self-employed, persons who work on gas fittings in domestic and commercial premises must be members of a body approved by the HSE, and such an approved body must operate within criteria prescribed by the HSE within the framework of the Regulations.
Under the scheme, gas fitters must have their competence assessed and must possess a certificate of the accredited certification body, which a consumer can ask to see prior to work being undertaken.
CORGI is empowered to appoint inspectors to undertake work and site inspections and to take appropriate action to secure safety of a dangerous or potentially dangerous installation is identified.

Council of Ministers (2B2) The final decision-making body in the European Union comprising representatives of member states.

counselling (1E3) The provision of advice, support and assistance to individuals with health, vocational or personal problems.

coupler In tubular steel scaffolding, a scaffold connection created by a steel tube joining the ends of two scaffold tubes together, usually by a form of friction grip.

court hierarchy (England and Wales) (1B1) There are two distinct systems – those courts dealing with criminal offences and those dealing with civil matters. Criminal cases are heard in the Magistrates' Court and may be referred to the Crown Court.

The Magistrates' Court is the lowest of the courts; magistrates hear cases and sentence for less serious offences, and also hold preliminary examinations into other offences to ascertain whether the prosecution can show a prima facie case, as a result of which the accused may be committed for trial at a higher court.

The Crown Court deals with serious criminal charges and appeals from Magistrates' Courts.

Civil cases are heard in the County Courts and the High Court of Justice. County Courts deal with simpler civil cases, e.g. civil claims in respect of negligence. The High Court deals with more important civil matters.

The three divisions of the High Court are Queen's Bench (contracts and torts), Chancery (land, wills, partnerships, companies, etc.) and Family. Queen's Bench has a supervisory function over the lower courts and tribunals.

The Court of Appeal has two divisions: the Civil Division hears appeals from County Courts and the High Court Criminal Division hears appeals from Crown Courts.

The House of Lords hears appeals on important legal matters only from the Court of Appeal and, in some cases, from the High Court.

The European Court of Justice is the supreme law court and its decisions are enforceable through the network of courts and tribunals in all member states.

creep (2D4) The progressive deformation of a material with time at constant stress, the deformation being irreversible. Creep can take place at both high and low temperatures. However, it is principally associated with static structures and high temperatures, as with steam boiler parts and turbine blades. Under these conditions, metals yield very slowly, so that over a period of time they stretch and may ultimately fracture.

crime An offence against the state.

criminal court procedure (1B1) A defendant, one who is accused of a crime, can be brought before the court either by:

- a summons, namely a written order, signed by a magistrate, ordering that person to appear before a certain court on a certain date at a specified time to answer the charge or accusation, which is the principal means; or
- a warrant, namely a written authority, issued by a magistrate, addressed to a

constable directing him or her to carry out some specified act, namely to arrest the person named in the warrant, and to bring that person before the court.

A summons is served by post or by hand, usually by a police constable.
Proceedings for health and safety offences can be commenced only by an inspector or with the consent of the Director of Public Prosecutions.
Offences under the HSWA can be prosecuted as:

- triable only summarily – Magistrates' Court;
- triable either way – Magistrates' Court or Crown Court;
- triable only on indictment – Crown Court.

Prosecution procedure takes the following steps:

- Information is laid before the magistrate or clerk to the justices.
- The examining magistrate decides whether there is a charge to be answered.
- The magistrate decides whether the charge will be tried summarily or on indictment:
  - if tried summarily, the decision is explained to the defendant, who consents or otherwise to the summary trial;
  - if tried on indictment, the Magistrates' Court proceedings are known as committal proceedings.

Bail may be granted to the accused pending trial, depending upon the severity of the offence, and the accused can appeal to a judge in chambers where bail is refused.
Not all witnesses need attend a Magistrates' Court, as a written statement may be accepted as evidence.
Two sides are involved – the prosecution and the defence.
In Magistrates' Courts, the procedure is governed by the Magistrates' Courts Act 1980, whereby the charge is read out and the defendant asked to state whether he or she is guilty or not guilty. Where the defendant denies the charge, witnesses may be called and the evidence of witnesses challenged through the process of cross-examination. After the cross-examination stage, the magistrates make a decision and, upon conviction, the defendant will be sentenced.
Crown Court procedure is as follows:

- There is an arraignment, namely the calling of the accused, the reading of the indictment and the calling on the accused to plead.
- Where a not guilty plea is entered, a jury will be sworn in; the procedure takes the form of reading the prosecution charge and the giving of evidence, followed by a defence submission or 'no case to answer'.
- Where 'no case to answer' is accepted by the judge, the jury is directed to find the accused not guilty; where this submission is rejected the trial continues, with defendant and witnesses giving evidence.
- On completion of the evidence and cross-examination stage, the judge sums up and the jury considers its verdict.

- The judge may accept a majority verdict of not less than 10:2; where a jury cannot agree a verdict, the judge may make a direction to return a not guilty verdict.
- On a guilty verdict, the judge can hear evidence of both good and bad character which he may or may not take into consideration; the judge may impose a sentence or discharge the accused, i.e. fine and/or imprisonment, suspended sentence, conditional discharge, absolute discharge, etc.

criminal liability (1B1) This refers to the responsibilities of persons under statutes and the penalties that can be imposed by the criminal courts.
A crime or a breach of criminal law is an offence against the state. The burden of proving a criminal charge beyond reasonable doubt rests with the prosecution. Where a person is found guilty, a court will impose some form of punishment, such as a fine, or imprisonment, or both.
Compensation may be ordered by a court to be given to a person to cover personal injury and damage to property.
Cases are heard initially in the Magistrates' Court, but may be referred to the Crown Court for trial before a judge and jury.

crisis A situation where something happens that requires major decisions to be made quickly.

critical incidents A technique of carrying out task analysis by compiling instances that are critical for undertaking the task satisfactorily and safely, as distinguished from those representing work that can be done by anyone and which are not important in determining satisfactory completion of the task.

Cross-Sectional Survey (2E5) The study of groups of people at a particular stage of development, different groups being contrasted and compared at each stage of development.

culture The customs, habits and traditions that characterize an organization or social group. It includes the attitudes and beliefs that the organization or group has about, for instance, profitability, commitment, politics and, in some cases, health and safety.

cyanosis The bluish appearance of the skin caused by a deficiency of oxygen in the blood, commonly arising from exposure to toxic substances or through simple anoxia or asphyxiation.

cybernetics The study of the flow of information around an animate or inanimate system in which the information is used by the system as a means of controlling itself.

D

daily personal noise exposure The level of daily personal noise exposure of an employee ascertained in accordance with Part 1 of the Schedule to the Regulations, but taking no account of the effect of any personal ear protector used.
(Noise at Work Regulations 1989)

damage Death or personal injury or any loss of or damage to any property (including land).
(Consumer Protection Act 1987)

Damage Control Damage control is a feature of the Total Loss Control philosophy. In a loss control programme, damage control covers the control of accidents which result in damage to property and plant and which might, conceivably, eventually cause death or injury. By reporting, recording and investigating damage type accidents, more serious accidents involving people can be eliminated.
Damage control entails damage reporting by employees, recording of same, analysis of causes and costing.

damages Civil liability consists of an award of damages for injury, disease or death at work in circumstances disclosing a breach of common *law* and/or statutory *duty* (normally negligence) on the part of an employer/occupier of premises, arising out of and in the course of employment.
General damages relate to losses incurred after the hearing of an action, namely actual and probable loss of future earnings following an accident.
Special damages relate to quantifiable losses incurred before the hearing of the case, and consist mainly of medical expenses and loss of earnings.
In the case of fatal injury, compensation for death negligently caused is payable under the Fatal Accidents Act 1976, and a fixed lump sum is payable under the Administration of Justices Act 1982 in respect of bereavement.

danger Liability or exposure to harm; a thing that causes peril.
When applied to machinery in motion it is a situation in which there is a reasonably foreseeable risk of injury from mechanical hazards associated with contact with it or being trapped between the machinery and any material in or at the machinery, or any fixed structure. Or being struck by, or entangled in or by any material in motion in the machinery or being struck by parts of the machinery ejected from it, or being struck by material ejected from the machinery.
(BS 5304: *Safeguarding of Machinery*)

dangerous for the enviroment Substances which, were they to enter into the environment, would or might present an immediate or delayed danger for one or more components of the environment.

[Chemicals (Hazard Information and Packaging for Supply) Regulations 1994]

dangerous occurrence An event listed in Schedule 2 of the Reporting of Injuries, Diseases and Dangerous Occurrences Regulations (RIDDOR) 1995. It is an event with particularly significant potential for death and/or major injury, such as the collapse or overturning of lifting machinery, unintentional explosions, gassing accidents and boiler explosions.
Under RIDDOR, dangerous occurrences are classified in five groups – general, those relating to mines, those relating to quarries, those relating to relevant transport systems and those in respect of offshore workplaces.

danger zone Any zone in and around machinery in which a person is exposed to a risk to health or safety from contact with a dangerous part of machinery or a rotating stock bar.
(Provision and Use of Work Equipment Regulations 1998)

deadstop device A form of safety trip device, incorporating a trip bar or arm, an electrical switch and brake, located in close proximity to a danger point on a machine. When the trip arm (e.g. the telescopic arm featured on a fixed vertical drill) is touched, the machine motor automatically cuts out, bringing the machine to rest.

decibel A tenth of a bel, the unit of sound intensity. It is essentially the logarithm of a ratio of pressures or energies.

defect There is a defect in a product if the safety of the product is not of the kind one would generally be entitled to expect.
(Consumer Protection Act 1987)

defences (negligence and breach of statutory duty) (1B1, 2B3) Where presented with a civil claim, a defendant may deny liability on the following grounds:

- that the duty alleged to have been breached was never owed to the claimant in the first place;
- that the nature of the duty was different to that pleaded by the defendant – that the duty was complied with;
- that the breach of duty did not lead to the injury, damage or loss in question;
- that the claimant was partly to blame, i.e. was guilty to contributory negligence, which resulted in the injury, damage or loss.

deflagration (2C3) The chemical reaction of a material in which the reaction front advances into the unreacted material at less than sonic velocity.

deformation When a metal is put under stress, it will firstly deform but subsequently return to the original form when the stress is relaxed. Where the elastic limit is exceeded, however, plastic deformation occurs and permanent set is produced where the metal does not return to its original shape.

delegated (subordinate) legislation Legislation made by the Queen in Council, or a Minister or Secretary of State under powers contained in a parent statute. It is exercised through Statutory Instruments (SI) and formerly Statutory Rules or Orders (SR&Os).

demolition (2C4) The process of dismantling a structure and the most dangerous of construction operations.

Table D.1 Guide to typical methods of demolition[1]

Type of structure	Type of construction	Method of demolition[2]			
		Detached building, isolated site	**Detached building, confined site**	**Attached building, isolated site**	**Attached building, confined site**
Small and medium two-storey buildings	Load-bearing walls	ABCDM	ABDM	ABDM	ADM
Large buildings	Load-bearing walls	ABDM	ABDM	ABDM	AD
three storeys and over	Load-bearing walls with wrought iron and cast iron members	ABDM	AM	AM	AM
Framed structures	Structural steel	ACM	AM	AM	AM
	In situ reinforced concrete	ADM	ADM	ADM	AM
	Precast reinforced concrete	ADM	ADM	ADM	AM
	Prestressed reinforced concrete				
	Composite (structural steel and reinforced concrete)	ADM	ADM	ADM	AM
	Timber	ABCDM	ABDM	ABDM	ABDM
Independent cantilevers (canopies, balconies and staircases)		ADM	ADM	ADM	ADM
Bridges		ABCDM	ABCDM	AM	AM
Masonry arches		ACDM	ACDM	ACDM	ACDM
Chimneys	Brick or masonry	ACD	A	ACD	A
	Steel	AC	A	A	A
	In situ and precast reinforced concrete	AD	A	AD	A
	Reinforced plastics	AC	A	A	A
Spires		ACD	A	A	A
Pylons and masts		AC	A	A	A
Petroleum tanks (underground)					
Above ground storage tanks					
Chemical works and similar establishments					
Basements					
Special structures					

[1]This table is a general guide to the methods of demolition usually adopted in particular circumstances. In addition, subject to local restraints, explosives may be used by experienced personnel in many of the circumstances listed. This table should be read in conjunction with the main text. The indication of a particular method does not necessarily preclude the use of another method, or the use of several methods in combination.

[2]Key: A denotes hand demolition, B denotes mechanical demolition by pusher arm, C denotes mechanical demolition by deliberate collapse, D denotes mechanical demolition by demolition ball and M denotes demolition by other mechanical means excluding wire pulling.

Demolition may be undertaken by a number of techniques, including manual techniques, and a range of mechanical techniques, e.g. by the use of a pusher arm, by deliberate collapse, such as wire pulling, and by the use of a demolition ball. See Table D.1.

demonstration An active form of training where the trainer, by actually performing what needs to be done and how to do it and, with associated explanations, indicates why, when and where it is to be done. Demonstration frequently accompanies other forms of training.

depression A mood, characterized by feelings of dejection and gloom, and other combinations, such as feelings of hopelessness, futility and guilt.
A sadness which has lost its relationship to the logical progression of events (David Viscott, American psychiatrist).

dermatitic substance (1E1) A hazardous substance, exposure to which causes dermatitis.
The term 'dermatitic' is not classified as a Category of Danger in Schedule 1 of the Chemicals (Hazard Information and Packaging for Supply) Regulations 1994.

dermatitis Dermatitis (inflammation of the skin) is the most common occupational disease. Prevention is largely aimed at, or associated with, improvements in personal hygiene.
'Non-infective dermatitis' is classified as a *prescribed disease* in relation to dust, liquid, vapour or other skin irritants.
Most cases of dermatitis are either:

- endogenous – controlled by factors within the person, e.g. dermatitis may be stress induced;
- exogenous – controlled by factors from outside the person, e.g. exposure to plants, organic substances, acids, etc.

Causes of dermatitis are associated with:

- primary irritants, which cause dermatitis at the site of contact if permitted to act for a sufficient length of time in sufficient concentration, e.g. solvents, strong acids;
- secondary cutaneous sensitizers, such as plants, rubber and many chemical compounds, which do not cause skin changes at first contact, but effect a specific sensitization, resulting in future dermatitis at the slightest exposure.

The stages of dermatitis are redness of the skin, swelling, blistering, cracking, scaling and crusting.
Dermatitis is frequently a first indication of an unsatisfactory degree of exposure to a toxic material.

descriptive statistics Statistical measures that summarize the characteristics of a frequency distribution or the relationship between two or more distributions.

designer Any person who carries on a trade, business or other undertaking in connection with which they:

(a) prepare a design; or
(b) arrange for any person under their control (including where the person is an employer, or an employee of theirs) to prepare a design relating to a structure or part of a structure.

[Construction (Design and Management) Regulations 1994]

design stress A safety consideration in the design of structures with a view to ensuring that materials comprising a structure are never subjected to stress that could result in failure of the structure.

$$\text{Design stress} = \frac{\text{Ultimate tensile stress}}{\text{Factor of Safety}}$$

The Factor of Safety may vary between 4 and 10.

deterministic problems In maintenance management, these are problems where no uncertainty is associated with the timing or consequence of the maintenance action.

detonation (2C3) The extremely rapid chemical reaction of a material in which the reaction front advances into the reacted material at greater than sonic velocity. The resultant blast wave is initially characterized by a very high peak pressure acting over a very short time.

detoxification (2A3) The process in the human body when decomposition of toxic substances occurs to produce harmless substances which are eventually eliminated from the body.

diffusion The projection of light in many directions but with no directional predominance.

diffusion sampling (passive sampling) A form of air sampling where the contaminant passes over the sampling train in natural air currents and diffuses into a chamber containing a sensitive absorbent material.

digestive system (1E1) Digestion is the process of breaking down food into substances that can be absorbed and used by the body to provide energy, growth and repair of human tissue.
The digestive system principally comprises the mouth, oesophagus, stomach, small and large intestines. Most food absorption takes place in the small intestine (ileum).

dilution A control strategy used, particularly, in the transportation of hazardous wastes, where the substance is diluted with water to the point where it is no danger or less dangerous.

dilution ventilation (1E3) In certain situations it may not be possible to use a

local exhaust ventilation system to remove airborne contaminants. Where the quantity of contaminant is small, uniformly evolved and of low toxicity, it may be possible to dilute the contaminant by inducing large volumes of air to flow through the contaminated region.
Dilution ventilation is most successfully used to control vapours from low toxicity solvents, but is seldom successfully applied to dust and fumes.

direct current Electric current which flows in one direction only.

disability glare The visually disabling effect produced by bright bare lamps directly in the line of sight. (Dazzle)

disablement benefit A state benefit for loss of amenity paid to persons who may not be entitled to statutory sick pay following injury at work.

disclosure of information Section 28 of the HSWA requires that no person shall disclose any information obtained by him or her as a result of the exercise of any power conferred by sections 14 or 20 (including, in particular, any information with respect to any trade secret obtained by him or her in any premises entered by him or her by virtue of any such power) except:

- for the purpose of his functions;
- for the purposes of any legal proceedings, investigation or inquiry, for the purpose of a report of any such proceedings or inquiry or of a special report made by virtue of section 14; or
- with the relevant consent.

Information must not normally be disclosed except with the consent of the person providing it.
Disclosure may be made in certain cases:

- for the purposes of any legal proceedings, investigation or inquiry held at the request of the HSC;
- with the relevant consent;
- for providing employees or their representatives with health and safety-related information.

discomfort glare A form of glare caused by too much contrast of brightness between an object and its background.

discrimination (2B4) A process whereby an individual, over a period of time, learns to select a particular positive stimulus, as opposed to an alternative stimulus. Generally, responses to the positive stimulus are reinforced, whereas responses to the negative stimulus are extinguished.

display Any means of presenting information to a person.

display screen equipment (1D1) Any alphanumeric or graphic display screen, regardless of the display process involved.
[Health and Safety (Display Screen Equipment) Regulations 1992]

The regulations apply only to 'users' and 'operators' of display screen equipment:

- a 'user' means an employee who habitually uses display screen equipment as a significant part of his normal work;
- an 'operator' means a self-employed person who habitually uses display screen equipment as a significant part of his normal work.

display screen equipment risk analysis Under the Health and Safety (Display Screen Equipment) Regulations 1992, employers and self-employed persons must undertake a suitable and sufficient analysis of those workstations used by users and operators.
The risk analysis must take into account criteria stated in the Schedule to the regulations thus:

- equipment – use of equipment, display screen, keyboard, work desk or work surface, work chair;
- environment – space requirements, lighting, reflections and glare, noise, heat, radiation, humidity;
- interface between computer and operator/user.

distance guard A guard which does not completely enclose a danger point or area but which puts it out of normal reach.
(BS 5304: *Safeguarding of Machinery*)

distribution (lighting) The way in which light is spread by a luminaire. To give evenly distributed illuminance, luminaires should be spaced regularly. Evenness of illuminance depends upon the ratio between the height of the luminaire above the working position and the spacing of fittings. (Distribution Ratio)

domestic client A client for whom a project is carried out, not being a project carried out in connection with the carrying on by the client of a trade, business or other undertaking (whether for profit or not).
[Construction (Design and Management) Regulations 1994]

Domino Theory (Heinrich) (1A4) This theory is well known as one of the great theories of accident causation. Heinrich likened the sequence of accidents to five standing dominoes. If the first domino falls, the remaining dominoes will also fall in a particular sequence. However, if one of the dominoes is removed, the sequence of falling is broken and the end result, the last domino falling, cannot take place.
Heinrich said that a preventable accident is one of five factors in a sequence that results in injury. The injury is invariably caused by the accident and the accident, in turn, is always a result of the factor that immediately precedes it.
Heinrich's five factors or events are:

1. Ancestry and social environment.
2. Fault of the person.
3. Unsafe act and/or mechanical or physical hazard.

4. Accident.
5. Injury.

Bird and Loftus subsequently extended Heinrich's Domino Theory to take into account the influence of management in the accident sequence. They proposed an alternative sequence of events, thus:

1. Lack of control by management, permitting
2. Basic causes, i.e. personal and job factors, leading to
3. Immediate causes, e.g. substandard practices, conditions or errors, which are the direct cause of
4. The accident, which results in
5. A loss, e.g. negligible, minor, serious or catastrophic.

dose (2E3) The level of environmental contamination or offending agent related to the duration of exposure to same.

Dose = Level of environmental contamination × Duration of exposure

The term is used in the case of physical stressors (e.g. noise), chemical stressors (e.g. gases) and biological stressors (e.g. bacteria).

dose–effect relationship (2E3) Estimation of the relationship between the specific dose of a contaminant and its effects on the human body is based on the degree of association existing, firstly, between an indicator of dose (i.e. urine, faeces, blood, saliva) and, secondly, an indicator of effect on the body (e.g. respiratory difficulties, unconsciousness, headaches).
The study of this relationship will show the particular concentration of a toxic substance at which the indicator of effect exceeds the value currently accepted as 'normal'.

dose equivalent (2E3) The biological effects of exposure to alpha, beta, gamma and neutron radiation vary considerably. The dose equivalent takes into account a modifying or weighting factor to the absorbed dose due to this variation. Dose equivalent, which is measured in sieverts (Sv), indicates a measure of potential biological damage resulting from exposure to radiation.

dose limit (radiation) This term refers to the sum of all exposures from external irradiations or internal irradiation based upon exposure over a calendar year, with the exception of women of reproductive capacity and pregnant women.
Dose limits are classified according to age and employment status, and are set for the whole body (uniform radiation) and for individual organs and tissues (non-uniform radiation except the lens of the eye).

dose–response relationship (2E3) Consideration of threshold limits of exposure or dose, which most people can tolerate without either short-term or long-term damage to their health, is a basic feature of the prevention and control of occupational diseases.
For many chemicals commonly used, it is possible to establish a relationship or

link between the dose received and the body's response (e.g. coughing, lachrymation), a characteristic known as the 'dose–response relationship'.
Where dose is plotted against response in a graphical form, with many dusts, for instance, the response is directly proportional to the dose. In the case of other environmental contaminants, the dose–response curve remains at a level of no response at a point greater than zero on the dose axis. This point of cut-off identifies the threshold dose. After reaching the threshold dose, the body's response rises dramatically.

double-barrelled action An employee is entitled to sue his or her employer for damages for injury resulting from a breach of a duty at common law and a statutory duty in certain situations (the double-barrelled action). In this case, an injured employee sues separately, though simultaneously, for a breach of both duties by the employer.
This form of action can be traced back to Kilgollan v. Cooke & Co.Ltd. (1956), a case involving the machinery fencing provisions of the Factories Act 1937.

dry chemical powder appliance A form of fire appliance, using a dry chemical powder, which breaks the chemical chain of reignition. There are two types – stored pressure and gas cartridge.

dual circuit interlocking A form of electrical interlocking of a machinery guard used where direct mechanical linkage between a guard gate and the back-up exhaust or dump valve is impracticable. In this case, electrical actuation of the back-up is acceptable. The control and back-up limit switches are arranged in opposite modes, the control being negative and the back-up positive.

ductile failure (2D4) A form of failure occurring when a metal is stretched to the point where the yield stress has been exceeded over a large area. The metal reaches the plastic stage and failure by ductile fracture takes place.

ductility The property of a metal which enables it to be given a high level of mechanical deformation, especially by stretching, without cracking.

dust An aerosol composed of solid inanimate particles (International Labour Organization).
Dusts are solid airborne particles frequently created by processes such as grinding, milling, crushing, sanding and demolition. They may be prejudicial to health and a nuisance to local inhabitants.

dust explosion (2C1) Most particulates (i.e. dusts, powders) will form an explosive mixture in a particular combination with air. When exposed to a source of ignition, a dust explosion will occur.
Certain overriding conditions must be present, namely:

- the dust must be airborne in a concentration within its explosive range;
- the ignition source must be strong enough to initiate combustion, typically 50 mJ minimum;
- there must be sufficient oxygen in the air to support combustion.

duty holder (2C2) A term used in certain regulations, such as the Electricity at Work Regulations 1989 and the Construction (Design and Management) Regulations 1994, specifying classes of person on whom duties are imposed, e.g. employers, the self-employed, clients, contractors and planning supervisors.

duty of care (1B1) Employers must take reasonable care to protect their employees from the risk of foreseeable death, injury or disease at work. This common duty of care can extend to work on third party premises or where an employee may have been hired out to another employer.

E

ear defender/muff A form of hearing protection which covers the outer ear completely.

ear plug A form of hearing protection manufactured in glass down, plastic or rubber which is inserted at the entrance to the auditory canal. They may be the permanent or disposable type.

ear protection zone A demarcated area where employees must wear ear protection.

earth bond test A test carried out on portable electrical appliances. In this test a substantial test current, at approximately 25 amps, is applied via the earth pin of the plug to an earth test probe which should be connected by the user to any exposed metalwork on the casing of the unit under test. From this, the resistance of the earth bond is determined by the portable appliance testing (PAT) device.

earthing (1D4) This implies connection to the general mass of earth in such a manner as will ensure at all times an immediate discharge of electrical energy without danger.
Earthing, to give protection against indirect contact with electricity, can be achieved in a number of ways, including the connection of extraneous conductive parts of premises (radiators, taps, water pipes) to the main earthing terminal of an electrical installation.

earth leakage circuit breaker (residual current device) (1D4) A device, similar in appearance to an enclosed switch, which incorporates a mechanism that trips the switch from the ON to the OFF position if an excess of current flows in an electrical circuit. The circuit breaker must be of the type and rating for the particular circuit and appliance that it protects.

earth leakage test A further test of the insulation under its true working conditions carried out during the operation test of a portable electrical appliance. This test should further ensure that appliances are not responsible for nuisance tripping of residual current devices.

ear valve A form of hearing protection inserted into the auditory canal which permits normal conversation to take place whilst preventing hearing damage.

EC Declaration of Conformity (2D2) Information relating to equipment, such as lifting appliances and electrical equipment, which must be produced by a manufacturer and kept available for inspection by an enforcing authority, generally for a period of 10 years after manufacture of the equipment has ceased.
[Electrical Equipment (Safety) Regulations 1994]

A declaration which complies with Regulation 22 of the Supply of Machinery (Safety) Regulations 1992 and Regulation 8(2)(d) of the Lifts Regulations 1997. (Lifting Operations and Lifting Equipment Regulations 1998)

ego In psychoanalysis, a term referring to the self and to ways of behaving and thinking realistically. The ego directs motives into socially acceptable channels.

ejection A form of machinery hazard whereby particles or items are emitted or thrown off by a machine.

elasticity (2D4) The property of a material which permits strain to disappear on the removal of stress.
All materials have a specific elastic limit, below which the material will return to its original shape after the removal of the load.

elastic limit The stage in the application of stress to a material where the material fails to return to its original length.

electrical burns (1D4) Where a current passes through a conductor, heat is produced. Burns can be caused through contact with hot conductors or by a passage of a current through the body at the points of entry and exit. Electric arcing from short circuits may also cause burns.

electrical equipment This includes anything used, intended to be used or installed for use, to generate, provide, transmit, transform, rectify, convert, conduct, distribute, control, store, measures or use electrical energy.
(Electricity at Work Regulations 1989)

electrical interlocking A form of interlocking of machinery guards. There are four methods:

- control interlocking;
- power interlocking;
- control interlocking with back-up;
- dual circuit interlocking.

See individual entries.

electrical resistivity test (2D4) A form of non-destructive testing of materials which is based on the fact that the resistance between two wires can indicate flaws in the surface. This is based on the time taken for the current to pass around the flaw or crack being greater than that if no flaw or crack existed.

(electrical) system This means any electrical system in which all the electrical equipment is, or may be, electrically connected to a common source of electrical energy, and includes such source and such equipment.
(Electricity at Work Regulations 1989)

electric arc (1D4) Electrical burns can be caused by hot arcs when short circuits arise in equipment. Arcing can be generated by relatively low voltages, but with high-energy power sources. Generally, the higher the voltage of the power supply,

the greater the gap an arc can cross. Arcing commonly occurs during electrical testing to measure voltage. An arc may be struck when a conductor is brought close to an exposed live conductor which is carrying high voltage.
Protection measures include:

- ensuring that live conductors are insulated or protected;
- limiting the amount of exposed metalwork which is brought into contact with a live conductor.

electric shock (1D4, 2C2) The effect produced on the body, and in particular the central nervous system, by an electric current passing through it. A common cause of death is ventricular fibrillation (spasm) of the heart muscle, which can occur at 0.05 amps.

electrode A conductor which conveys electric current directly into the body of an electric furnace, vat or other electrical apparatus.

electromagnetic radiation A general term referring to a variety of physical changes in the environment, including light, radio waves, X rays and cosmic rays. See Figure E.1.

Radiation	Frequency (Hz)	Wavelength	Energy	Radiation sources
Gamma	10^{21}	Short	High	Cosmic sources
X-ray	10^{18}	↑	↓	Atoms struck by high-energy particles
Ultraviolet light				Excited gases
Visible light	10^{15}			Hot bodies
Infrared	10^{12}–10^{14}			Hot bodies
Microwaves	10^{9}			Microwave generator
Radio waves	-10^{6}	Long	Low	Radio transmitter

Figure E.1 The electromagnetic spectrum.

electronic gas detector A device for detecting and measuring flammable gases based on their reaction with an electric filament. The resulting combustion and rise in temperature can be detected electronically and displayed as a concentration in air.

elimination (1E3) A prevention strategy in the use of hazardous substances whereby substances no longer in use, or which can be replaced by less hazardous substances, are eliminated from an organization's inventory and stock of substances.

Emergency Action Code A Code which informs the emergency services of the action to be taken in the event of an accident. Codes are approved by the HSC for dangerous substances specified in the Approved List and must appear on a hazard warning panel fitted to vehicles carrying such substances.

emergency device A device which stops a machine (e.g. a conveyor) immediately in the event of an emergency. It may take the form of a trip wire and/or emergency stop button.

emergency escape respirator A type of respirator using a chemical filter and intended solely for short-term use to enable people to escape from dangerous atmospheres in an emergency situation.

emergency evacuation procedure (1C5) The Management of Health and Safety at Work Regulations 1999 require employers to 'establish and where necessary give effect to appropriate procedures to be followed in the event of serious and imminent danger to persons at work'.

Emergency procedures should be based on a 'worst case scenario' basis, e.g. a major escalating fire, collapse of a structure.

Emergency procedures take account of four phases of an emergency:

- preliminary action;
- action when the emergency is imminent;
- action during the emergency; and
- ending the emergency.

emergency lighting This form of lighting is provided to ensure safety when a normal lighting installation fails.

Standby lighting enables essential work to continue, the illuminance required depending upon the nature of the work. It may be between 5 and 100 per cent of the illuminance provided by the normal lighting installation.

Escape lighting enables a building to be evacuated safely, and may take the form of battery- or generator-powered installations.

emergency maintenance (2A3) Maintenance which it is necessary to put in hand immediately to avoid serious consequences.

employers' liability The liability owed by an employer towards his employers, both under criminal law and civil law.

employment business A business (whether or not carried out with a view to profit and whether or not carried out in conjunction with any other business) that supplies people (other than seafarers) who are employed in it to work for and under the control of others in any capacity.

(Management of Health and Safety at Work Regulations 1999)

Employment Medical Advisory Service (1E3) This service operates under the Employment Medical Advisory Service Act 1972 and by virtue of Part II of the HSWA. It advises the Secretary of State of Employment, the HSC, government bodies and others concerned with the health and safety of employed persons on health-related issues at work.

The EMAS provides medical assistance and appoints registered medical practitioners as Employment Medical Advisers.

Enabling Act An Act of Parliament which provides for, or enables, a Minister or

Secretary of State to produce subordinate or delegated legislation, e.g. HSWA 1974.

Energy Trace and Barrier Analysis (ETBA) An analytical process forming part of the Management Oversight and Risk Tree (MORT) system, ETBA is based on the view that energy is necessary to do work. On this basis, energy must be controlled, and uncontrolled energy that flows in the absence of barriers can cause accidents. The simple energy–barriers–targets concept is expanded with the details of specific situations to answer the question 'What happened?' in an accident.

Enforcement Notice Where the enforcing authority is of the opinion that a person carrying on a prescribed process under an authorization is contravening any condition of that authorization, or is likely to contravene any such conditions, the authority may serve an enforcement notice.
An enforcement notice shall:

- state that the authority is of the said opinion;
- specify the matters constituting the contravention or the matters, making it likely that the contravention will arise, as the case may be;
- specify the steps that must be taken to remedy the contravention or to remedy the matters that make it likely that the contravention will arise, whichever is relevant in the circumstances;
- specify the period within which these steps must be taken.

(Environmental Protection Act 1990)

enforcement procedures Enforcement of health and safety legislation is undertaken by Inspectors appointed under Section 20 of HSWA and authorized by written warrant from the enforcing authority, e.g. the HSE, local authority.
Enforcing authorities are:

- Health and Safety Executive (HSE) – factories, construction, nuclear installations, etc. inspectorates;
- local authorities – principally by environmental health officers;
- fire authorities – for certain matters, e.g. fire certification.

entanglement A machinery hazard associated with the entanglement of hair, limbs and clothing in rotating machinery parts. See Figure E.2.

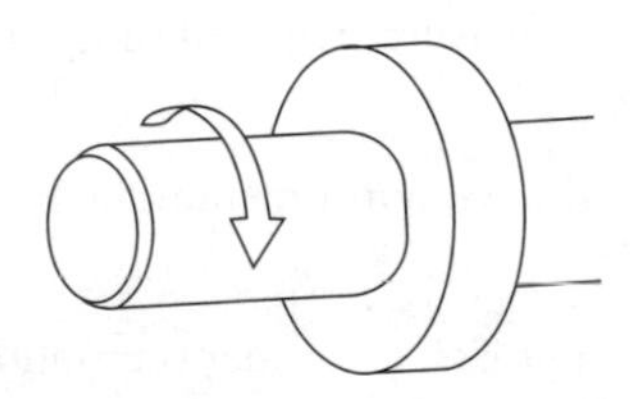

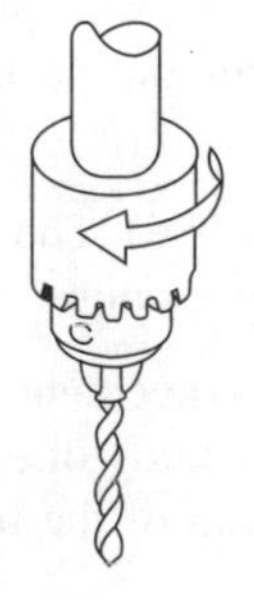

Figure E.2 Entanglement risks.

environment This consists of all or any of the following media, namely, the air, water and land; the medium of air includes the air within buildings and the air within other natural or man-made structures above or below ground. (Environmental Protection Act 1990)

environmental factors Factors within the workplace, including:

- temperature, lighting, ventilation and humidity;
- the control of noise and vibration;
- the prevention/control of dust and fume emissions;

which may have a direct effect on the health of employees and other persons. In addition to the duties on employers under the HSWA for ensuring an environment which is, so far as is reasonably practicable, safe and without risks to health, the Workplace (Health, Safety and Welfare) Regulations 1992 place absolute requirements on employers to ensure a safe and healthy working environment. In particular:

- every room where persons work shall have sufficient floor area, height and unoccupied space for the purposes of health, safety and welfare;
- structural features should ensure the safety of floors and traffic routes, stairs, ladders and catwalks, windows, doors, gates and walls, and external areas, traffic routes and approach roads;
- no waste material or refuse should be allowed to accumulate within a working area and an adequate supply of containers should be provided at convenient points, together with an external storage and disposal area;
- a reasonable temperature must be provided in all workplaces;
- suitable and sufficient lighting must be provided throughout a workplace, together with emergency lighting where there may be a risk in the event of failure of the main lighting system;
- ventilation systems must be effective and suitable, taking into account both comfort standards and the need for the removal of dust, gases and other airborne contaminants;
- humidity levels must be controlled to maintain maximum comfort.

environmental monitoring This is concerned with the identification, measurement and evaluation of a range of physical, chemical and biological agents (e.g. noise, dust, bacteria, gases) in the workplace atmosphere. It may be undertaken on an intermittent and/or continuous basis using a range of measurement and sampling equipment.

environmental pollution (2C6) The pollution of the three environmental media – land, water and air – by a range of agents.

Epidemiological Approach (accident causation) An approach to accident causation based on the formalized study of epidemics. One theory suggests an accident is caused by the conjunction of the following:

- Host factors – factors related to the accident victim, e.g. age, sex.
- The agent – the object that directly gave rise to the accident, either an abnormal energy exchange that produced the injury or the specific types of damage.
- The environment, which is subdivided into:
  – physical, e.g. seasons, geography;
  – biological, e.g. poisons, toxins;
  – socio-economical – the fact that some individuals are more susceptible to accidents than others.

epidemiology The study of the distribution of diseases in different groups of the population.
The study of the distribution and determinants of disease frequency.

ergonomics (1E1) Ergonomics can be defined in several ways:

- the scientific study of work;
- human factors engineering;
- the study of the man–machine interface;
- the scientific study of the interrelationships between people and their work.

Ergonomics takes into account:

- the human system;
- environmental factors;
- the man–machine interface; and
- the total working system.

esteem needs According to Maslow, the need for prestige, self-respect and success, a group of motives in the hierarchy of human motivation.

European Commission (2B2) One of the four official bodies in the European Union (EU), it performs a civil service-like function and is headed by a body of Commissioners from member states. It is empowered to take action against any member state not complying with EU legislation. The Commission can make proposals for future legislation to the Council of Ministers.

European Court of Justice (2B2) The supreme law court within the European Union. Cases can only be brought before this court by organizations or individuals representing organizations.

European Directive (2B2) The Council of the European Commission can issue Directives, which provide for harmonization of the laws of the member states, including those dealing with occupational health and safety. Directives impose a duty on each member state to produce legislation to conform to the Directive and to enforce such legislation.
Directives are legally binding on all governments.
Framework Directives set out overall objectives. More detailed matters are subsequently dealt with in Daughter Directives.

European Parliament (2B2) This consists of members (MEPs) who represent constituencies in their individual member states. The mail function of the European Parliament is to debate EU proposals. It must be consulted on all proposed European legislation.

evacuation The process of moving people from a building or area to a place of safety and/or identified assembly area.

Event Tree Analysis (2A1) A safety management system, similar to fault tree analysis in approach, this technique works from a selected 'initiating event', such as the failure of an electrical or pressure system. It is a systematic representation of all the states of a processing operation conditional to the specific initiating event and relevant for a certain type of outcome.

excess current protection (2C2) Where excessive current flows in a circuit, the resultant overheating could cause fire.
Circuits must, therefore, be protected by the use of fuses and circuit breakers. However, these forms of protection may not give total protection against electric shock. Where this risk exists, a residual current circuit breaker should be installed.

exhaustion The third stage in the General Adaptation Syndrome (Selye) in which a person is no longer capable of enduring stress.

exothermic reaction That feature of a combustion reaction whereby energy is emitted.

expansion heat detector This type of heat detector consists of a bimetallic strip which expands in a circular mode under the influence of heat, closing off an electrical circuit.
A second form involves the expansion of liquids, as with a quartzite bulb used in sprinkler systems. Here the final 'valve' consists of a quartzite bulb filled with a liquid. On contact with fire, the liquid expands. At a predetermined temperature the bulb bursts and water is released.

explosion detection The potential for explosion is assessed by the measurement of the concentration of a flammable substance in air on a percentage basis and within the flammable range of the substance. Explosimeters will detect the presence of a specified gas or vapour and the concentration. They indicate a dangerous condition, generally at one third of the lower flammable limit.
The installation of explosion detection equipment to initiate inerting and isolating arrangements, coupled with automatic plant shut-down, is a standard means of protecting major installations against dust explosions.

explosion relief/vent (2C1) This is a form of precaution directed at reducing the damage following an explosion in a dust-producing plant. It generally takes the form of a lightweight panel installed at the top of, for instance, an evaporator drier, silo, cyclone or elevator.
The size of the explosion relief is related to the volume of the installation and its mechanical strength. See also ***explosion venting***.

explosion suppression (2C1) Explosion suppression devices are a standard feature of major installations where there may be a risk of dust explosions.

Dust explosions do not take place instantaneously but develop due to a continuous build-up of pressure to the maximum pressure. This continuing rise in pressure can be suppressed by an explosion suppression system, which arrests the rise in pressure before it reaches a dangerous level.

An explosion suppression system consists of a detector, an electrical power unit and a number of suppressors. See Figure E.3.

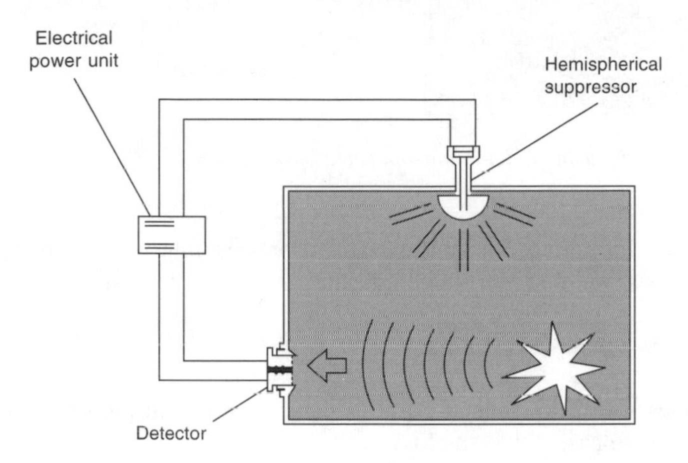

Figure E.3 Explosion suppression using an explosion detector, electrical power unit and a hemispherical suppressor.

explosion venting (2C3) Dust explosions create a shock wave which can cause death, major injury and substantial damage to property.

Explosion venting systems are commonly installed in industrial premises (such as mills) and in certain plants (such as elevators, pneumatic handling systems and ducts), where there may be a risk of dust explosion. Such systems incorporate 'blow out' panels which, when an explosion takes place, allow the shock wave to vent in a safe direction, usually to the external air.

Careful selection of the size of venting panels is essential. See Figure E.4.

explosive Solid, liquid, pasty or gelatinous substances and preparations which may react exothermically without atmospheric oxygen, thereby quickly evolving gases, and which under defined test conditions detonate, quickly deflagrate or, upon heating, quickly explode when partially confined.

[Chemicals (Hazard Information and Packaging for Supply) Regulations 1994]

extremely flammable Liquid substances and preparations having an extremely low flashpoint and a low boiling point, and gaseous substances and preparations which are flammable in contact with air at ambient temperature and pressure.

[Chemicals (Hazard Information and Packaging for Supply) Regulations 1994]

eyebolt An item of lifting tackle, fixed to a load which may be heavy and concentrated, prior to lifting.

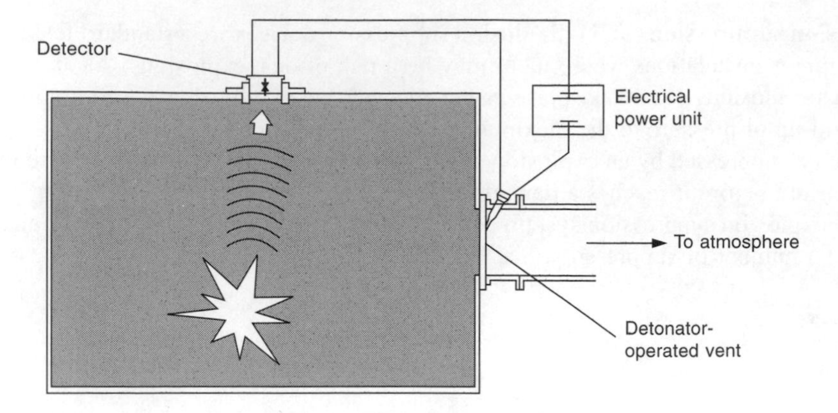

Figure E.4 Explosion venting using a detonator-operated bursting disc.

eye protection (1E4) Eye protection is designed to prevent injury from flying particles, dust, fumes, welding glare and splashes from hazardous substances, such as acids.

Eye protection takes a number of forms:

- safety spectacles – with toughened glass or plastic lenses, plastic or metal frames and, in some cases, side shields;
- safety goggles – which incorporate a cup surrounding each;
- eye and face shields – which can be hand-held, fixed to a safety helmet or strapped to the head.

F

face mask A simple form of respiratory protection incorporating a thin metal shield which holds a filtering medium against the nose and mouth, with the intention of preventing coarse nuisance dust particles and paint spray entering same. Face masks have a very limited application and should not be used to protect people against toxic airborne substances.

face velocity (1E3) The rate of air movement at the commencement of a local exhaust ventilation system, e.g. a partial enclosure or receptor hood.

Factor of Safety A term used in the design of safe structures and in the materials used for such structures.

$$\text{Factor of Safety} = \frac{\text{Ultimate Stress}}{\text{Working Stress}}$$

The Factor of Safety for different materials (e.g. steel, concrete) varies immensely.

fail safe A design feature of machinery whereby any failure in, or interruption of, the power supply to a safeguard will result in the prompt stopping or, where appropriate, stopping and reversal of the movement of the dangerous parts before injury can occur, or the safeguard remaining in position to prevent access to the danger point or area.

Failure Mode and Effect Analysis (2A1) This technique is based on identifying the possible failure modes of each component of a system and predicting the consequences of that failure.

falsework Any structure used to support a permanent structure whilst it is not self-supporting.
(BS 5975: 1982)

fatigue (2A2) The effects of prolonged work activity and/or lack of sleep; a feeling of being permanently tired.

Fault Tree Analysis (2A1) A form of safety management technique which begins with the consideration of a chosen 'top event', such as the collapse of a crane, and then assesses the combination of failures and conditions which could cause this event to arise.
The technique is widely used in quantitative risk analysis, particularly where control over process controls is critical to meeting safety standards.

feedback The situation in which some aspect of the output of an activity or event regulates or contributes to future inputs to the system, e.g. feedback from accidents may contribute to the design of safe systems of work.

feed check valve A non-return valve situated on a boiler shell or steam drum, generally just below water level, which prevents water escaping from the boiler should the pressure in the water feed line be less than that of the boiler.

fibrogenic dust Dusts which may cause fibrotic changes to lung tissue, e.g. cement dust, coal dust, silica and some metals.

film badge A personal radiation monitoring device or dosemeter incorporating a film sensitive to radiation housed in a specifically designed plastic casing. The badge, which is worn on the clothing of the person who may be exposed, incorporates windows of various materials which shield certain kinds of radiation but allow other kinds to pass through. Analysis of the film enables assessment of the various kinds of radiation to which a person may have been exposed over, normally, a 1-week period. The accumulated dose of radiation for each operator is recorded on this basis.

filtration device A form of personal sampling device which comprises a low-flow or constant-flow sample pump, which is motor-generated from a rechargeable battery, and a sampling head. The head incorporates a specific filter and is attached close to the operator's breathing zone. The quantity of dangerous material to which the wearer has been exposed, such as dust, can be determined gravimetrically, i.e. weighing the filter before and after collection.

filtration methods (gravimetric sampling) (1E2) Techniques commonly used in dust sampling, whereby dust particles are filtered out of the air flow and the amount collected on the filtration medium, a filter paper, weighed. Where the rate of air flow and sampling time are known, the total volume of air that has passed through the filter can be evaluated. From this air volume and the weight of dust collected, the mean concentration of dust in air can be calculated.

final exit door A door giving direct access to a place of safety from the escape route.

fire Defined as:

- a spectacular example of a fast chemical reaction between a combustible substance and oxygen accompanied by the evolution of heat;
- a mixture in gaseous form of a combustible substance and oxygen with sufficient energy put into the mixture to start a fire;
- an unexpected combustion generating sufficient heat or smoke resulting in loss of life, damage to plant, equipment, goods and buildings.

fire alarm (1C5) A means for drawing the attention of people to the presence of a fire. The most effective fire alarm is the human voice.
Fire alarm systems can range from the use of a single hand bell to electronically actuated systems. Fire alarm systems are commonly linked with heat and/or

smoke detection systems, the detection system actuating the audible alarm and sprinkler systems.

fire appliance/extinguisher (1C5) Fire appliances are designed to be carried and operated by hand. They contain an extinguishing medium which can be expelled by the action of internal pressure and directed at a fire.
The pressure may be stored, or obtained by a chemical reaction, or by release of gas from a cartridge.
Fire appliances must be coloured red and display a distinguishing label thus:

Water	Red
Foam	Cream
Carbon dioxide	Black
Dry chemical powder	Blue

[Health and Safety (Safety Signs and Signals) Regulations 1996]

fireball The phenomenon which may occur as a result of the deflagration of a vapour cloud which does not result in a blast wave. The burning cloud may lift off the ground and form a mushroom-shaped cloud.

fire certificate (2C1) The Fire Precautions Act 1971 requires that a fire certificate be issued for certain classes of factory and commercial premises, based on the concept of 'designated use'.
A fire certificate specifies:

- the use or uses of the premises it covers;
- the means of escape in the case of fire indicated on a plan of the building;
- the means for ensuring the safety and effectiveness of the means of escape, such as smoke stop doorways, emergency lighting and direction signs;
- the means of fighting fire for the use of persons on the premises;
- the means of raising the alarms; and
- particulars of any explosive or highly flammable liquids stored and used on the premises.

A fire certificate may also impose requirements relating to:

- the maintenance of the means of escape and keeping it free from obstruction;
- the maintenance of the fire precautions;
- the training of people and the keeping of records;
- limitations on numbers of persons in the premises; and
- any other relevant fire precautions.

fire classification Fires are classified in four categories according to the fuel type and means of extinction.
Class A fires involve solid materials, normally of an organic nature, in which combustion occurs with the formation of glowing embers. Water, applied as a jet or spray, is the most effective means of extinction.

Class B fires involve liquids and liquefiable solids. Liquids fall into two groups:

- B(i) – miscible with water, e.g. methanol, acetone, acetic acid;
- B(ii) – immiscible with water, e.g. petrol, benzene, fats and waxes.

Foam, vaporizing liquid, carbon dioxide and dry powder can be used on both B(i) and B(ii) type fires. Water spray can be used on type B(i) but not on type B(ii) fires. There may also be some type of restriction on the type of foam which can be used, as some foams break down in contact with alcohols. In all cases, extinction is achieved mainly by smothering.
Class C fires involve gases or liquefied gases, e.g. methane, propane and butane. Both foam and dry powder can be used on small liquefied gas spillage fires, particularly when backed up by water to cool the leaking container.
Class D fires involve metals, e.g. magnesium and aluminium. They can only be extinguished by the use of dry powders which include talc, soda ash, limestone and dry sand. All extinguishers work by smothering.

fire detector (1C5) Fire can be detected by:

- sensing heat, i.e. the actual temperature or the rate of temperature rise;
- detecting the presence of smoke;
- detecting the presence of a flame or flames.

Heat detectors are of two kinds, i.e. fusion heat detectors and expansion heat detectors. See individual entries.
There are three kinds of smoke detector, i.e. ionizing detectors, light scatter detectors and light obscuration detectors. See individual entries.

fire door (2C1) A door forming part of a designated means of escape and/or giving direct access to the open air.

fire drill A planned or unplanned evacuation of a building or area in order to train and familiarize occupants with the means of escape.

fire exit A designated door giving direct access to the open air and forming part of a means of escape in the event of fire.

fire instructions A notice informing people of the action they should take on either hearing a fire alarm or discovering a fire. See Figure F.1.

fire point The lowest temperature at which the heat from combustion of a burning vapour is capable of producing sufficient vapour to sustain combustion.

fire precautions A fire authority may impose specific fire precautions (e.g. limiting the maximum number of people permitted in a building) as a condition or requirement of a fire certificate.

fire resistance The ability of a structural element of a building, such as a plastered 9-inch wall or a cavity wall, to withstand fire. Fire resistance of structural elements is expressed as the 'notional period of fire resistance' for which the particular structural element is capable of withstanding fire under specified test conditions.

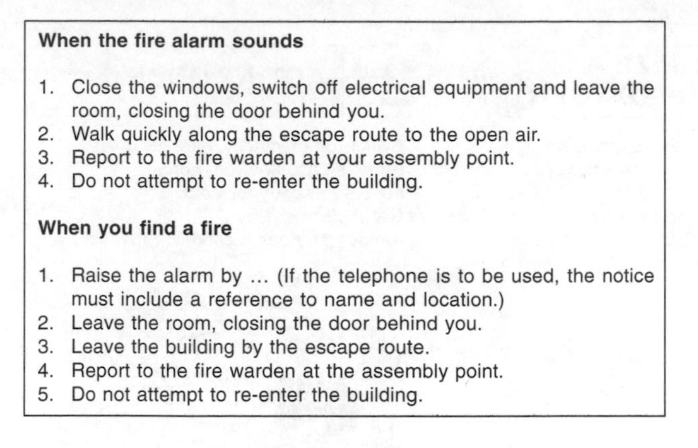
When the fire alarm sounds

1. Close the windows, switch off electrical equipment and leave the room, closing the door behind you.
2. Walk quickly along the escape route to the open air.
3. Report to the fire warden at your assembly point.
4. Do not attempt to re-enter the building.

When you find a fire

1. Raise the alarm by ... (If the telephone is to be used, the notice must include a reference to name and location.)
2. Leave the room, closing the door behind you.
3. Leave the building by the escape route.
4. Report to the fire warden at the assembly point.
5. Do not attempt to re-enter the building.

Figure F.1 Fire instruction notice.

fire resisting structure Any building, part of a building, structure, cabinet and enclosure which is constructed in conformity with a specification for fire-resisting structures, being a specification approved for the time being for the purposes of the Regulations by Certificate of the Chief Inspector.
(Highly Flammable Liquids and Liquefied Petroleum Gases Regulations 1972)

fire safety log book A form of record incorporating fire safety procedures, legal requirements, approved codes, guidance and management systems for fire prevention maintained by an organization.

fire triangle There are three elements required for the propagation of a fire – fuel, an ignition source and air containing sufficient oxygen to maintain combustion. The removal of any one of the three components prevents combustion commencing or being maintained. These three elements are commonly depicted as a triangle.

fire ventilation system A ventilation system that comes into operation in the event of fire in a building with the aim of removing smoke and heated gases quickly, thereby enabling evacuation and fire fighting operations to be undertaken more safely.

first action level A daily personal noise exposure of 85 dB(A).
(Noise at Work Regulations 1989)

First Aid (1C1) The skilled application of accepted principles of treatment on the occurrence of an accident or in the case of sudden illness, using facilities and materials available at the time.
The principal aims of first aid are:

- to sustain life;
- to prevent deterioration in an existing condition; and
- to promote recovery.

See Figure F.2.

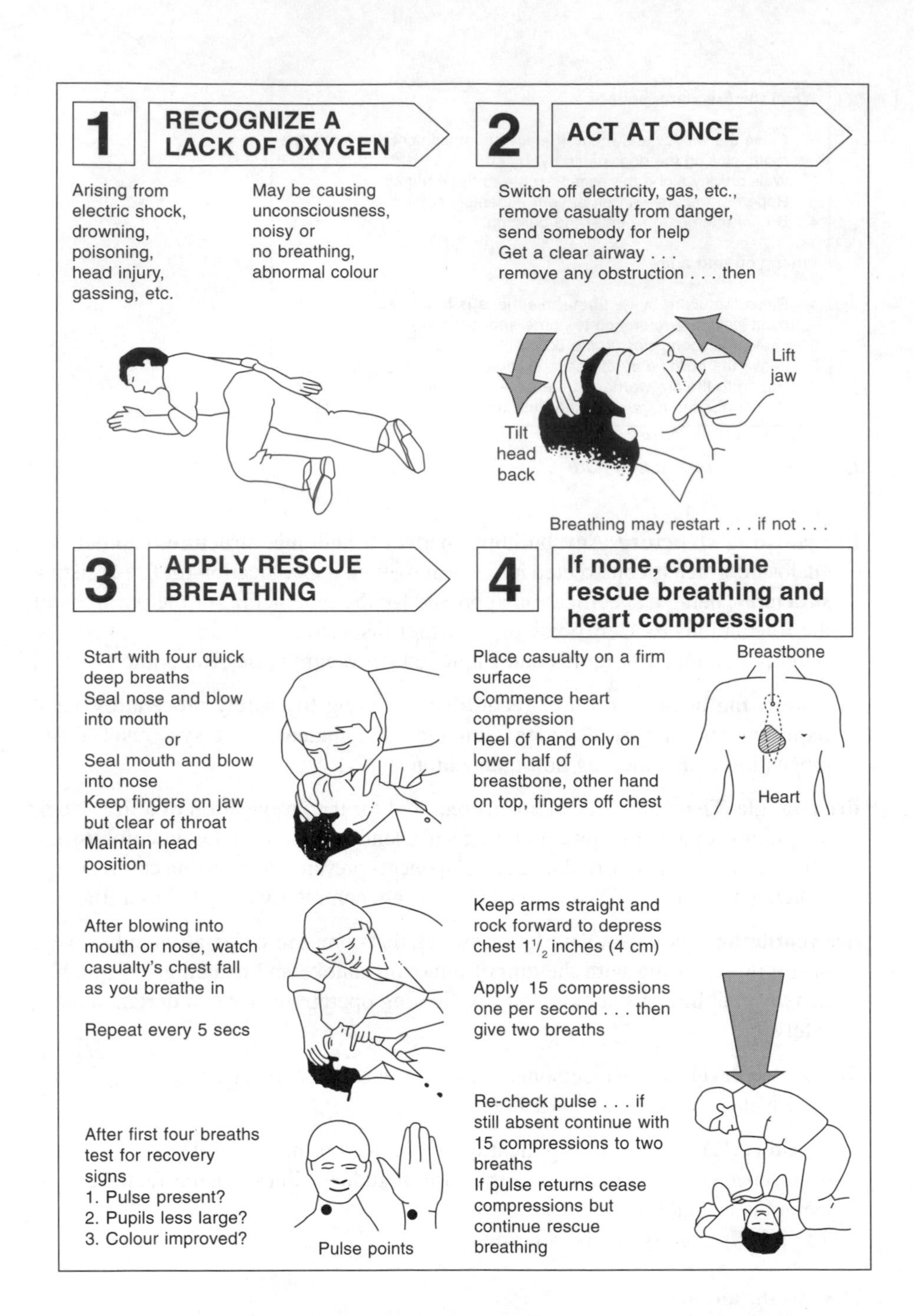

Figure F.2 First aid – resuscitation procedure.

Under the Health and Safety (First Aid) Regulations 1981, first aid means:

- in cases where a person will need help from a medical practitioner or nurse, treatment for the purpose of preserving life and minimizing the consequences of injury or illness until such help is obtained; and
- treatment of minor injuries which would otherwise receive no treatment or which do not need treatment by a medical practitioner or nurse.

fixed fire fighting equipment Various forms of permanent installation in buildings provided for dealing with fires, including hose reel installations, sprinkler systems and hydrants.

fixed guard A guard which has no moving parts associated with it, or dependent upon the mechanism of any machinery, and which, when in position, prevents access to the danger point or area.
(BS 5304: *Safeguarding of Machinery*)

fixed sequence robot A robot capable of performing successive steps of a given operation in a predetermined sequence, condition and position.

fixed term contract of employment A contact of employment for a specific term which is fixed in advance or which can be ascertained in advance by reference to relevant circumstance.
(Management of Health and Safety at Work Regulations 1999)

flame detector A form of heat detector operating on the principle of detecting either or both ultraviolet and infrared radiation emitted by a fire.

flame point The lowest temperature above the flash point at which a substance will ignite and continue to burn when the flame is removed.

flameproof equipment (2C2) Electrical equipment which can produce heat or sparking may be a source of ignition in flammable atmospheres. Where this risk may be present, flameproof equipment must be used.
Flameproofing of equipment should ensure that the equipment or apparatus is constructed to withstand any explosion, within the apparatus, which arises from the ignition of a flammable gas which could enter through the casing or other enclosure. To ensure adequate flameproofing, all flanges and other joints of the casing or enclosure of the apparatus must be so designed and constructed as to prevent any internal ignition of gas from passing out of the enclosure and igniting a surrounding flammable atmosphere.

flammability (1C5) The ability of a substance to burn naturally.

flammable (1C5) Liquid substances and preparations having a low flash point.
[Chemicals (Hazard Information and Packaging for Supply) Regulations 1994]

flammable atmosphere (1C5, 2C1) An atmosphere created by the presence of flammable gases, vapours and dust. In each, the concentration in air must be between the upper and lower explosive limits for ignition to be possible.

flammable gas detector A fire protection device which measures the concentration of flammable gas in an atmosphere, actuating an alarm when the concentration of flammable gas reaches a predetermined fraction of the lower explosive limit.

flammable zone (flammable atmosphere) (1C5, 2C1) A zone where a flammable atmosphere is, or may be, present. Flammable zones are categorized thus:

- Zone 0 – where a flammable atmosphere is always present.
- Zone 1 – where a flammable atmosphere is likely to be present.
- Zone 2 – where a flammable atmosphere is not likely to occur but if it does it will be for a short period of time.

flash point (1C5, 2C1) The minimum liquid temperature at which sufficient vapour is given off a liquid to form a mixture with air capable of ignition but not to continue burning under prescribed test conditions.
Flash point is determined by the use of the Abel apparatus, Abel–Pensky apparatus, Tag apparatus or Pensky–Martens apparatus.
Methods for determination of flash point are detailed in Schedule 1, Part III of the Chemicals (Hazard Information and Packaging for Supply) Regulations 1994.

flash test In the testing of portable electrical appliances, this is a more stringent test of the insulation which provides an indication of insulation defects developing in the appliance. It tests the insulation at a higher voltage, generally 1.5 kV for Class 1 appliances and 3 kV for Class 2 appliances, and from the portable appliance testing (PAT) device a leakage of current may be indicated.

flaw detection The technique for detecting internal flaws in materials by means of a range of radiographic techniques using gamma rays, neutrons and X-rays. Such tests detect the thickness of metal and the degree of fusion of welds on pipes and internal surfaces.

foam appliance These appliances take the form of a chemical foam fire extinguisher and a stored pressure foam extinguisher. See individual entries.

foot protection This includes safety boots, safety shoes, clogs, foundry boots, wellington boots, anti-static footwear and ankle protection. Foot protection should be waterproof, resistant to acids, alkalis, oils and other substances. It should incorporate, where appropriate, steel toecaps. In wet processes, safety wellingtons, incorporating a steel toecap, should be provided.

formal group A social group that has a relatively permanent structure of positions, jobs and roles.

form taken by a substance (1E1) The form taken by a substance hazardous to health (e.g. solid, gas, dust, vapour, fume, mist, fog or smoke) is significant in its potential for harm. See individual entries.

Fracture Mechanics (2D4) A branch of engineering science concerned with the study of material failures and factors which determine the probability of catastrophic failure of various structural components. The results of these studies can be used in the design of structures, machinery and lifting appliances.

Frequency Analysis (octave band analysis) (2E2) This is an important technique in, particularly, noise measurement.
Sound is distributed throughout the frequency spectrum, which is approximately 12 kHz in the case of human hearing. This is a very large scale for measurement purposes and, therefore, this scale is split into eighths or 'octaves' and the sound pressure measured at the geometric centre frequency of each octave band. An octave represents a doubling of frequency, so that the range 90–180 Hz is one octave, as is the range 2800–5600 Hz.
The standard range of octave bands is shown in Table F.1.

Table F.1 The standard range of octave bands

Limits of band (Hz)	Geometric centre frequency (Hz)
45–90	63
90–180	125
180–355	250
355–710	500
710–1400	1000
1400–2800	2000
2800–5600	4000
5600–11 200	8000

Frequency analysis is an important measure in analysing the sources of, particularly, machinery noise.

frequency distribution A set of measurements arranged from lowest to highest, or vice versa, and accompanied by a count (frequency) of the number of times each measurement or class of measurement occurs.

frequency (sound) (2E2) This is the number of complete pressure variations passing a fixed point per second. It is measured in hertz (Hz) and 1 Hz equals 1 cycle per second; 1 kHz equals 1000 cycles per second.
The frequency of a sound gives it its distinguishing character. High-frequency sound, such as that from a whistle, is perceived as high pitched, whereas low-frequency sound, such as that from a double bass, will sound low pitched. The more rapidly these vibrations occur, the higher the frequency, and vice versa.

fretting A form of wear in pin-jointed, press-fitted components, where there should be no movement, but where there is small relative motion between joints and parts. The surface finish is destroyed promoting fatigue and corrosion.

fume (1E1) Solid particles which usually form an oxide on contact with air.
Fume is created by processes involving the heating and melting of metals, such as welding, smelting, lead burning, soldering and arc air gouging, and can arise as a product of combustion.

fumigation An operation in which a substance is released into the atmosphere so as to form a gas to control or kill pests or other undesirable organisms.

(Control of Substances Hazardous to Health Regulations 1999)

fuse A form of electrical protection comprising a strip of metal of such size as would melt at a predetermined value of current flow. It is placed in an electrical circuit and, on melting, cuts off the electrical supply to the circuit.

fuse test In the testing of portable electrical appliances, a test to indicate the integrity of the fuse and that the appliance is switched on prior to other tests being undertaken.

fusible plug A plug of metal with a low melting point set in a boiler shell at low level. In the event of the water level falling and the boiler overheating, the plug melts, allowing the boiler water to escape and douse the fire.

fusion heat detector (1C5) This form of heat detector operates on the basis of a metal melting, completing an electrical circuit and releasing water.
In its simplest form, this detector consists of an electrical circuit incorporating a switch held in either the open or closed position by a piece of low-melting alloy. In the heat of the fire this alloy melts, the switch is released and the circuit conditions change.

G

galvanic skin response A change in the electrical resistance of the skin, occurring in emotion and certain other conditions.

gamma rays A form of ionizing radiation, these are very-high-energy electromagnetic waves that are emitted at the same time as alpha and beta particles. They are similar in nature to natural light but with very much shorter wavelengths.

gas (1E1) A formless fluid which normally seeks to fill the space into which it is liberated.

gas badge A form of personal sampling instrument designed to give an indication of exposure to specific gases. The badge, which contains a solid sorbent or a chemically impregnated carrier, is worn on the worker's clothing, the gas coming into contact with the badge by the process of diffusion. The results of exposure can be read directly by a colour change or specific analysis.

gas cartridge appliance This type of fire appliance operates on the basis of a plunger breaking a pressure cylinder.
In the case of a gas cartridge water extinguisher, carbon dioxide is contained in a small pressure cylinder, the seal being broken by the plunger. The gas released expels the water from the appliance.
With a gas cartridge foam extinguisher, a type of stored pressure appliance, the foam concentrate is expelled as foam as a result of breaking the seal on a carbon dioxide cartridge.

gas chromatography A technique for the analysis of mixtures of gases, whereby a sample of the airborne contaminant is injected into a carrier gas stream, which passes through a tube incorporating a stationary phase. Each component of the gas mixture distributes itself between the gas phase and the moving phase, thereby reducing the rate of travel of these components and allowing them to separate. After separation, each component of the mixture travels down the tube at a different speed and can be measured by a detector located at the end of the tube.

gas incident Any death or any major injury which has arisen out of or in connection with the gas distributed, filled, imported or supplied, as the case may be, by a conveyor of flammable gas through a fixed pipe distribution system, or a filler, importer or supplier (other than by means of retail trade) of a refillable container containing liquefied petroleum gas.
(Reporting of Injuries, Diseases and Dangerous Occurrences Regulations 1995)

gas transporter A person or organization who manages the flow of gas through pipeline systems.
[Gas Safety (Management) Regulations 1996]

Geiger–Muller counter A device for the detection and measurement of radioactivity. It can be used for monitoring background radiation to a very accurate degree, in particular low levels of radiation.

General Adaptation Syndrome One of the most well-known theories of the stress process developed by Dr Hans Selye (1936), showing how people adapt to threatening circumstances. This takes place in three defined stages:

1. The alarm reaction stage – typified by receiving some form of shock at the time when the body's defences are down, followed by counter shock, when the defences are raised, a 'call-to-arms' on the part of the body's defences.
2. The resistance stage – concerned with two responses whereby the body will either resist the stressor or adapt to the effects of the stressor.
3. The exhaustion stage – a situation where the stressor continues to act on the body, resulting in loss of the acquired adaptation and overloading of the body systems.

This syndrome is commonly known as the 'Flight or Fight Syndrome'. See Figure G.1.

general damages Damages arising from losses incurred after the hearing of a civil action, namely actual and probable loss of future earnings following an accident.

general purpose dust respirator A form of dust respirator incorporating an ori-nasal face mask and a particulate filter which will trap finely divided solid or liquid particles.

generic risk assessment A risk assessment produced once only for a given activity, type of workplace or specific work group. This is particularly appropriate where organizations run a range of similar workplaces in different locations (e.g. maintenance workshops), undertake activities which are standard (e.g. tyre fitting) and employ people to carry out the same type of work in different locations (e.g. postmen).
For generic assessments to be effective:

(a) 'worst case' situations must be considered; and
(b) provision should be made within the generic risk assessment to monitor implementation of the recommended preventive measures and controls which are relevant with a particular workplace, work activity or work group.

genetically modified organism An organism is genetically modified if any of the genes or other genetic material in the organism:

- has been modified by means of an artificial technique prescribed in regulations by the Secretary of State; or

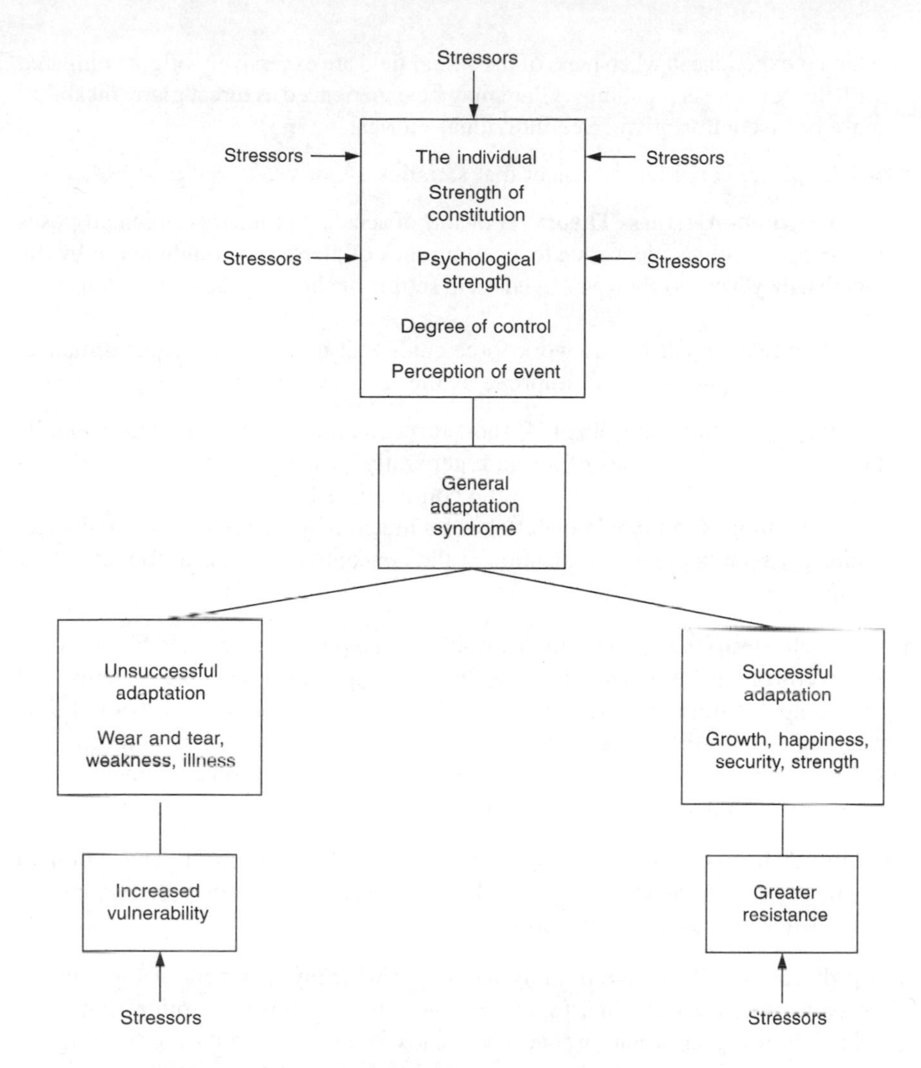

Figure G.1 Summary of the effects of stress on the individual.

- are inherited or otherwise derived, through any number of replications, from genes or other genetic material (from any source) which were so modified.
  (Environmental Protection Act, 1990)

genetic effect The effect following damage to the male or female reproductive cells as a result of exposure to radiation.

genetic mutation When genes or chromosomes are modified or altered, the hereditary characteristics may be altered so that characteristics will occur in the children which were not present in the parents.

glare (1E1, 2E3) The effect of light, which causes discomfort or impaired vision, which is experienced when parts of the visual field are excessively bright compared with the general surroundings. Glare may be experienced as direct glare, disability glare or discomfort glare. See individual entries.

goal The place, condition or object that satisfies a motive.

Goals–Freedom–Alertness Theory A theory of accident proneness which proposes that people have accidents due to a certain lack of alertness brought about by the fact that they have no choice or freedom in setting or choosing the goals established at work.

Freedom to set goals by the work force can result in high quality performance. This level of quality should improve as the level of alertness increases.

grab sampling (snap sampling) A short-term sampling technique which entails taking immediate samples of air and, generally, passing it through a particular chemical agent which responds to the contaminant being sampled.

Grab sampling is commonly undertaken by means of a stain tube detector device, which gives an immediate indication of the concentration in air of the offending agent.

gravimetric sampling (2E1) This form of air sampling may be undertaken on a short-term or long-term basis. The sampling technique entails drawing contaminated air through a filtration device and measuring the quantity of dust or other airborne contaminant by the process of weighing the filter paper before and after the collection period. This enables an assessment of the total dust or contaminant present over a period of time to be made.

grooving A form of mechanical corrosion in boilers arising from expansion and contraction at the junction of the firebox and boiler shell. It may be accelerated by a build-up of solids in the boiler.

group discussion An active training technique whereby members of a group are asked to discuss a particular topic or subject, with a view to learning from each other and using information that has already been imparted during training.

Group discussion can take a number of forms:

- directed discussion;
- developmental discussion;
- problem-solving discussion;
- syndicate exercises;
- group dynamics (T-groups);
- programmed instruction learning.

See individual entries.

Group Dynamics (T-groups) The study of the development and functioning of groups, with special reference to the interactions between groups and the patterns of relationships between individuals within the group.

A group discussion-based training technique in which situations develop, or are induced, in order to examine the behaviour and responses of the trainees. Group behaviour is also examined by the trainees.
The technique is effective in teaching trainees about their individual behaviour and the effects it has on others.

group norm A widely shared expectation or standard of behaviour amongst most members of a group.
The standards of behaviour set by a group and expected to be upheld by members of that group. Group norms are an important influencing factor in attitude and the changing of attitudes.

guard (1D2) A physical barrier to protect persons from danger.

guarding system (1D2) A combination of physical safeguards and safety devices applied to work equipment.

Guidance Notes (1B2) See ***HSE Guidance Notes***.

guided practice A training technique in which the trainee is required to perform the operation or procedure taught under controlled conditions. It may take the form of:

- independent practice;
- controlled practice;
- team performance;
- coach and pupil.

guided reading A training technique whereby a trainee is given prescribed material to read, and to subsequently comment on, in a structured situation.

H

haemolysis The action of haemolytic toxins which are capable of breaking down red blood cells.

haemolytic poison Poisons, such as arsine, which destroy the red cells of the blood, resulting in excretion of the destroyed red cells and a loss of their function.

half-life The time required for one half of a quantity of radionuclide to disintegrate, an indication of the time a radioactive substance will continue to emit radiation until decay takes place.

hand protection This includes gloves and gauntlets, principally concerned with protecting the hand from injuries, such as cuts, abrasions and burns, and those arising during manual handling operations and in extremes of temperature.

harm Harm to the health of living organisms or other interference with the ecological systems of which they form part and, in the case of man, includes offence caused to any of his senses or harm to his property.
(Environmental Protection Act 1990)

harmful Substances and preparations which may cause death or acute or chronic damage to health when inhaled, swallowed or absorbed via the skin.
[Chemicals (Hazard Information and Packaging for Supply) Regulations 1994]

hazard (1A1) Something with the potential to cause harm. This can include substances or machines, methods of work and other aspects of work organization.
The result of a departure from the normal situation, which has the potential to cause death, injury, damage or loss.

Hazard and Operability Studies (HAZOPS) (2A1) The application of a formal critical examination of the process and engineering intentions of new facilities to assess the hazard potential from incorrect operation or malfunction of individual items of equipment, and the consequential effects on the facility as a whole.
This technique is applied in the assessment of potential hazards from new installations and processes. HAZOPS is used extensively in high risk industries, such as the chemical industry, and in chemical engineering applications.

hazard effects (1A1) The anticipated effects arising from exposure to hazards, such as death, injury or disease.

hazard identification and evaluation (1A1) The identification of hazards and the evaluation of the risks arising from such hazards. It entails the identification of every form of hazard and devising means for eliminating, preventing or controlling exposure.

hazardous area classification (2C2) See *zoning*.

hazardous substances Hazardous substances and preparations are classified according to their category of danger under the Chemicals (Hazard Information and Packaging for Supply) (CHIP) Regulations 1994.
Classification is on the basis of:

- physico-chemical properties – explosive, oxidizing, extremely flammable, highly flammable, flammable;
- health effects – very toxic, toxic, harmful, corrosive, irritant, sensitizing, carcinogenic, mutagenic, toxic for reproduction;
- dangerous for the environment.

Certain substances may have a double classification, e.g. 'toxic' and 'flammable'. See Tables H.1 and H.2, and individual entries.

hazard rating A technique of hazard evaluation taking into account the frequency of the hazard arising multiplied by three factors, namely the severity of injury arising, the maximum possible loss (MPL) and the probability of the hazard arising. MPL and probability are measured on a scale from 1 to 50.

$$\text{Hazard Rating} = \text{Frequency} \times (\text{Severity} + \text{MPL} + \text{Probability})$$

Other techniques for rating hazards are used in addition to the technique outlined above.

hazard survey A review of all processes, methods and systems of work, taking into account:

- the exposure of employees and others, e.g. members of the public to risks;
- the value of the workplace, plant, machinery and equipment in economic and financial terms;
- the value of the process with regard to its interdependence with other processes;
- any inherent fire, explosion or other risk associated with any stage of the manufacturing operation.

head protection This form of protection includes:

- safety helmets – to protect the head from falling objects and overhead hazards;
- industrial scalp protectors (bump caps) – to protect against striking fixed obstacles, scalping or entanglement;
- caps and hair nets – to prevent the hair from coming into contact with moving machinery or parts of machinery.

Health and Safety at Work etc. Act 1974 – Duties (1B2) This is the principal statute dealing with occupational health and safety. It covers all persons at work except domestic workers in private employment and extends to the prevention of risks to the health and safety of the general public.
The principal requirements are:

Table H.1 Hazardous substances classification–Part I. Categories of danger

Category of danger	Property (1)	Symbol letter
PHYSICO-CHEMICAL PROPERTIES		
Explosive	Solid, liquid, pasty or gelatinous substances and preparations which may react exothermically without atmospheric oxygen thereby quickly evolving gases, and which under defined test conditions detonate, quickly deflagrate or upon heating explode when partially confined.	E
Oxidizing	Substances and preparations which give rise to an exothermic reaction in contact with other substances, particularly flammable substances.	O
Extremely flammable	Liquid substances and preparations having an extremely low flash point and a low boiling point and gaseous substances and preparations which are flammable in contact with air at ambient temperature and pressure.	F+
Highly flammable	The following substances and preparations, namely: (a) substances and preparations which may become hot and finally catch fire in contact with air at ambient temperature without any application of energy; (b) solid substances and preparations which may readily catch fire after brief contact with a source of ignition and which continue to burn or to be consumed after removal of the source of ignition; (c) liquid substances and preparations having a very low flash point; (d) substances and preparations which, in contact with water or damp air, evolve highly flammable gases in dangerous quantities.[2]	F
Flammable	Liquid substances and preparations having a None low flash point.	
HEALTH EFFECTS		
Very toxic	Substances and preparations which in *very low quantities* can cause death or acute or chronic damage to health when inhaled, swallowed or absorbed via the skin.	T+
Toxic	Substances and preparations which in *low quantities* can cause death or acute or chronic damage to health when inhaled, swallowed or absorbed via the skin.	T
Harmful	Substances and preparations which may cause death or acute or chronic damage to health when inhaled, swallowed or absorbed via the skin.	Xn
Corrosive	Substances and preparations which may, on contact, with living tissues, *destroy* them.	C

Category of danger	Property (1)	Symbol letter
Irritant	Non-corrosive substances and preparations which, through immediate, prolonged or repeated contact with the skin or mucous membrane, may cause *inflammation*.	Xi
Sensitizing	Substances and preparations which, if they are inhaled or if they penetrate the skin, are capable of eliciting a reaction by *hypersensitization* such that on further exposure to the substance or preparation, characteristic adverse effects are produced.	
Sensitizing by inhalation		Xn
Sensitizing by skin contact		Xi
Carcino-genic[3]	Substances and preparations which, if they are inhaled or ingested or if they penetrate the skin, may induce *cancer* or increase its incidence.	
Category 1		T
Category 2		T
Category 3		Xn
Mutagenic[3]	Substances and preparations which, if they are inhaled or ingested or if they penetrate the skin, may induce *heritable genetic defects* or increase their incidence.	
Category 1		T
Category 2		T
Category 3		Xn
Toxic for repro-duction[3]	Substances and preparations which, if they are inhaled or ingested or if they penetrate the skin, may produce or increase the incidence of *non-heritable adverse effects* in the progeny and/or an impairment of male or female reproductive functions or capacity.	
Category 1		T
Category 2		T
Category 3		Xn
Dangerous for the environment[4]	Substances which, were they to enter into the environment, would or might present an immediate or delayed danger for one or more components of the environment.	N

[1]As further described in the Approved Classification and Labelling Guide.

[2]Preparations packed in aerosol dispensers shall be classified as flammable in accordance with the additional criteria set out in Part II of this Schedule.

[3]The categories are specified in the Approved Classification and Labelling Guide.

[4](a) In certain cases specified in the Approved Supply List and in the Approved Classification and Labelling Guide substances classified as dangerous for the environment do not require to be labelled with the symbol for this category of danger.(b) This category of danger does not apply to preparations.

Table H.2 Hazardous substances–hazard warning symbols, as required by the CHIP 2 Regulations

Category of danger	Symbol letter	Indication of danger	Symbol
Explosive	E	Explosive	
Oxidizing	O	Oxidizing	
Extremely flammable	F+	Extremely flammable	
Highly flammable	F	Highly flammable	
Flammable			
Very toxic (very poisonous)	T+	Very toxic	
Toxic	T	Toxic	
Harmful	Xn	Harmful	
Corrosive	C	Corrosive	
Irritant	XI	Irritant	
Carcinogenic Categories 1 and 2	T	Toxic	
Category 3 (may cause cancer)	Xn	Harmful	
Mutagenic Category 1	T	Toxic	
Categories 2 and 3 (may cause mutation)	Xn	Harmful	
Teratogenic Category 1	T	Toxic	
Category 2 (may cause monstrosities)	Xn	Harmful	
Dangerous for the environment	N	Dangerous for the environment	

- Duties on employers to ensure, so far as is reasonably practicable:
  - the health, safety and welfare of all employees at work; and more specifically
  - to provide and maintain plant and systems of work that are safe and without risks to health;
  - to make arrangements for ensuring safety and the absence of risks to health in connection with the use, handling, storage and transport of articles and substances;
  - to provide information, instruction, training and supervision to ensure health and safety;
  - to maintain a safe workplace, with safe access to and egress from it;
  - to provide and maintain a safe working environment and adequate arrangements for welfare at work;
  - to prepare and revise as necessary a Statement of Health and Safety Policy;
  - to consult with trade union appointed safety representatives;
  - not to charge employees in respect of anything done or provided to ensure legal compliance (Section 2).
- Duty on employers to conduct their undertaking in such a way as to ensure, so far as is reasonably practicable, the health and safety of persons not in their employment (Section 3).
- Duty on persons who control premises to ensure, so far as is reasonably practicable, safety of premises, safe access and egress, and plant or substances used on the premises are safe and without risks to health (Section 4).
- Duties on persons who design, manufacture, import or supply articles or substances for use at work:
  - to ensure, so far as is reasonably practicable, safety of articles and substances when properly used;
  - to carry out testing and examination;
  - to provide adequate information about the safe use of articles and substances (Section 6).
- Duties on employees while at work:
  - to take reasonable care for the safety of themselves and other persons who may be affected by their acts or omissions at work;
  - to co-operate with the employer to enable compliance with the relevant statutory provisions (Section 7).
- Duty on any person not to intentionally or recklessly interfere with or misuse anything provided in the interests of health, safety and welfare in pursuance of any of the relevant statutory provisions (Section 8).

Health and Safety Commission A corporate body consisting of a Chairman, appointed by the Secretary of State for Employment, and not less than six nor more than nine other members.

health and safety culture (2B1) See *safety culture*.

Health and Safety Executive A corporate body consisting of three persons, of whom one shall be appointed by the Health and Safety Commission with the approval of the Secretary of State for Employment to be the Director. The others

are appointed by the Commission with the approval of the Secretary of State after consultation with the Director.
The UK national health and safety enforcement authority.

health and safety file A document containing information for the client or user of a building on the risks that may be present during maintenance, repair or renovation. [Construction (Design and Management) Regulations 1994]

health and safety plan A document required to be produced in conjunction with project work.
At pre-tender stage it must include a general description of the work, timings, details of risks to workers, general information for the principal contractor and welfare arrangements.
At the construction stage it must include arrangements for the health and safety of all those affected by the construction work, arrangements for management of the work and for monitoring legal compliance, together with information about welfare arrangements.
[Construction (Design and Management) Regulations 1994]

health and safety policy (1B3) See ***Statement of Health and Safety Policy***.

health promotion (1E3) All the measures taken by an employee to promote a healthy lifestyle amongst employees, including measures to increase awareness to the risks to health associated with alcohol, drugs and smoking.

health protection (1E4) All the measures taken by an employer to protect the health of employees and others whilst at work, including health surveillance, the provision of information, instruction and training, specific measures for protecting pregnant workers and young persons, and the control of health risks. See ***health surveillance***.

health risk assessment Where there may be a risk of exposure of employees to a substance hazardous to health an employer must make a suitable and sufficient assessment of the risks created by that work to the health of those employees and the steps that need to be taken to meet the requirements of the regulations.
An assessment of the risks created by any work should involve:

(a) Consideration of:
  (i) which substances or types of substance (including biological agents) employees are liable to be exposed to (taking into account the consequences of possible failure of any control measure provided to meet the requirements of regulation 7);
  (ii) what effects those substances can have on the body;
  (iii) where the substances are likely to be present and in what form;
  (iv) the ways in which and the extent to which any groups of employees or other persons could potentially be exposed, taking into account the nature of the work and process, and any reasonably foreseeable deterioration in, or failure of, any control measure provided for the purposes of regulation 7.

(b) An estimate of exposure, taking into account engineering measures and systems of work currently employed for controlling potential exposure.

(c) Where valid standards exist, representing adequate control, comparison of the estimate with those standards.

(Control of Substances Hazardous to Health Regulations 1999 and ACOP)

health surveillance (1E3) The specific health examination at a predetermined frequency of those at risk of developing further ill health or disability (e.g. employees exposed to chemical hazards) and those actually or potentially at risk by virtue of the type of work they undertake during their employment (e.g. radiation workers).

heat (1E1) A form of energy producing a sensation of hotness and transmitted by conduction, convection and radiation. See ***heat transmission***.

heat detector (1C5) A device installed in premises to detect heat which may arise from a fire. There are two main forms of heat detector, fusion heat detectors and expansion heat detectors. See individual entries.

Heat Stress Index (2E3) Whilst there are no specific heat exposure limits in the UK, threshold limits for permissible heat exposure (Indices of Thermal Stress) are used in the USA. The most commonly used index of thermal stress is that based on physiological observations and related to wet bulb globe temperature shown with a whirling hygrometer.
The following equations are used to calculate the wet bulb globe temperature (WBGT) values:

Outdoor work with solar load

$$WBGT = 0.7WB + 0.2GT + 0.1DB$$

Indoor work, or outdoor work, with no solar load

$$WGBT = 0.7WB + 0.3GT$$

where WB = natural wet bulb temperature, DB = dry bulb temperature and GT = globe thermometer temperature.
The natural wet bulb temperature is that record from the sling hydrometer without any rotation of the sling. After calculating the WBGT, the temperature in degrees Celsius is compared with recommended limits of work : rest schedules.
Workers should not be permitted to continue their work when their core temperature reaches 38°C.

heat transmission (1C5, 2C1) Heat may be transmitted by convection, conduction, radiation and direct burning.
Convection is a process of circulation of heat in liquids and gases by convection currents.
Conduction is a process of heat transmission mainly in solids. Certain substances are classed as good conductors of heat, such as metals, whereas others, such as wood, are bad conductors.
Radiation is heat transmission in air or gases where a body causes to be emitted, or radiated, into the air, rays or waves travelling in straight lines, in all directions

from the heat source. The radiation of heat is subject to the Inverse Square Law.

helmet-contained positive pressure respirator A form of respiratory protection which provides protection for the head, eyes, face and lungs, together with a high degree of comfort. The helmet incorporates a high-efficiency axial fan at the rear, which draws dust-laden air through a coarse filtration medium followed by a fine filter bag. The filtered air provides a cool air stream over the facial area and is finally exhausted at the bottom of the visor. Electrical power is supplied from a rechargeable battery pack clipped to the belt. See Figure H.1.

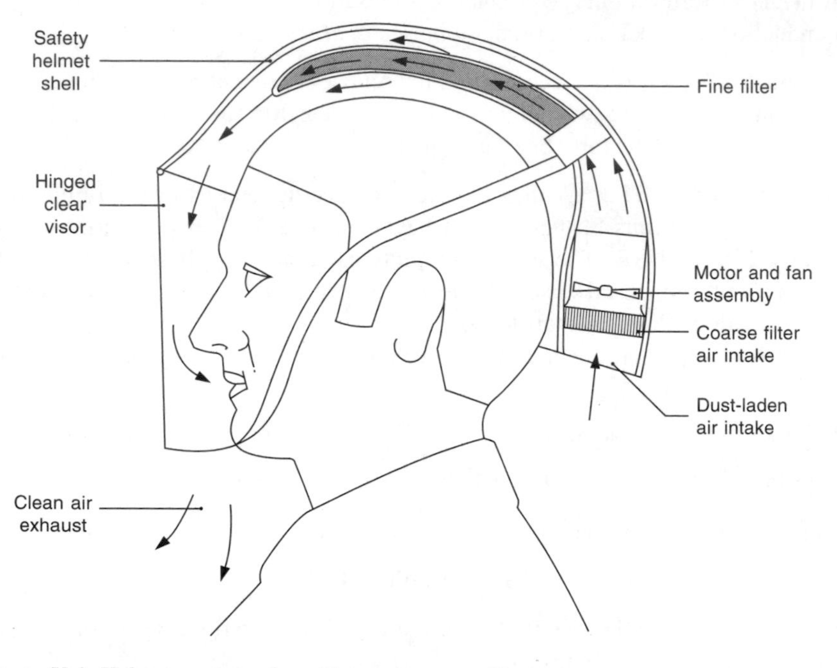

Figure H.1 Helmet-contained positive pressure respirator.

hernia A protrusion of an organ from one compartment of the body into another, e.g. a loop of intestine into the groin or through the frontal abdominal wall. Both the above forms of hernia can result from incorrect manual handling technique and particularly from the adoption of bent back stances, which produce compression of the abdomen and intestines.

hertz A unit of frequency equivalent to 1 cycle per second (Hz).

hierarchy of duties Statutory duties give rise to criminal liability and there are three distinct levels of duty.
Absolute or strict requirements (e.g. where the risk of injury or disease is inevitable if requirements are not followed); duties are qualified by 'shall' or 'must'.
'Practicable' requirements imply that if, 'in the light of current knowledge or

invention' or 'in the light of the current state of the art' it is feasible to comply with the requirements, irrespective of cost or sacrifice, then such requirements must be complied with (Schwalb v. Fass H. & Son, 1946).
'Reasonably practicable' is a narrower term than physically possible and implies that a computation must be made in which the quantum of risk is placed on one side of the scale and the sacrifice involved in carrying out the measures necessary for averting the risk is placed on the other side.
If it can be shown that there is gross disproportion between the above factors (i.e. the risk is insignificant in relation to the sacrifice), then the defendant discharges the onus on himself (Edwards v. National Coal Board, 1949).

high and low water alarm A safety device installed in a steam boiler which actuates an alarm when an incorrect level of water is present in the boiler, i.e. high water or low water.

High Court of Justice The senior court dealing with civil matters and comprising three divisions:

- Queen's Bench – contracts and torts;
- Chancery – land, wills, partnerships and companies;
- Family.

The Queen's Bench division hears appeals on matters of law:

- from the Magistrates' Courts and the Crown Court on the procedure called 'case stated';
- from some tribunals.

highly flammable Substances and preparations which may become hot and finally catch fire in contact with air at ambient temperature without any application of energy.
Solid substances and preparations which may readily catch fire after brief contact with a source of ignition and which continue to burn or to be consumed after removal of the source of ignition.
Liquid substances and preparations having a very low flash point.
Substances and preparations which, in contact with water or damp air, evolve highly flammable gases in dangerous quantities.
[Chemicals (Hazard Information and Packaging for Supply) Regulations 1994]

highly flammable liquid These include both liquefied flammable gas, although not aqueous ammonia, and liquefied petroleum gases, and therefore includes any liquid, liquid solution, emulsion or suspension which:

(a) gives off a flammable vapour at a temperature less than 32°C when tested in the manner set out in Schedule 1 to the regulations (closed cup flash point determination test); and
(b) supports combustion when tested in a manner set out in Schedule 2 (combustibility test).

(Highly Flammable Liquids and Liquefied Petroleum Gases Regulations 1972)

high voltage A high voltage is a voltage which exceeds 650 volts.
Persons working on high voltage installations must be competent and the work should be subject to a Permit to Work system.

histogram A frequency distribution represented by bars whose heights vary with the frequencies of the scores or classes of scores.

hoist (1D1, 1D3) A device for raising people and goods, taking the form of either a cantilevered or an enclosed structure.
There are three types of traction:

- rope and friction winch (for goods only);
- rope and geared drive;
- rack and pinion drive.

Goods hoists must be fitted with a safety device, such as holdback gears, in the event of rope failure, overrun trip devices, specific guards on the hoist machinery and landing gates, which must be securely fastened in the closed position during operation.
Passenger hoists must incorporate operating controls from inside the cage and interlock devices on the doors to prevent opening whilst in motion or away from a landing.

homeostasis The tendency of the human body to maintain a balance among internal physiological conditions, such as temperature, sugar level, oxygen level and mineral level.

Hooke's law This law states that, within the elastic limit of a material, stress is proportional to strain.

$$\text{Stress} = \text{Constant} \times \text{Strain}$$

$$\frac{\text{Stress}}{\text{Strain}} = \text{Constant}$$

hoop stress An induced stress set up in the curved surface of a cylinder by the pressure on the inner surface. Where the hoop stress is exceeded, the cylinder will split across its diameter.

hose reel A form of fixed fire fighting installation consisting of a coil of 25 mm i.d. flexible hose directly connected to a rising main. The complete fixed installation, with the hose reel as the terminal point, consists of either a wet or dry rising main and a landing valve or fire hydrant. The main consists of a heavy quality wrought steel pipe of not less than 100 mm i.d.

hot work The use of a range of equipment which may produce direct flames, heat, sparks and arcing, and involving processes such as welding, cutting, brazing, soldering and the boiling of bitumen.

HSE Guidance Notes (1B2) The Health and Safety Executive issues Guidance Notes which have no legal status and are purely of an advisory nature, to supplement information in Regulations and Approved Codes of Practice.
Guidance Notes are issued in six specific series:

- General Safety;
- Chemical Safety;
- Environmental Hygiene;
- Medical;
- Plant and Machinery;
- Health and Safety Guidance.

human engineering A field of specialization concerned with the design of machinery and equipment, and tasks performed in the operation of same.

human error (2A2) Limitations in human capacity to perceive, attend to, remember, process and act on information are all relevant in the context of human error. Human error is associated with lapses of attention, mistaken actions, misperceptions, mistaken priorities and, in some cases, wilfulness.

Human Factors (1A2) A term used to cover a range of issues, including:

- the perceptual, physical and mental capabilities of people and the interaction of individuals with their job and working environment;
- the influence of equipment and system design on human performance; and
- the organizational characteristics which influence safety-related behaviour.

These are affected by:

- the system for communication within the organization; and
- the training systems and procedures in operation.

All of these are directed at preventing human error.

Human Reliability Assessment (2A2) A technique which includes the identification of all the points in a sequence of operations at which incorrect human action, or the failure to act ('sins of omission'), may lead to adverse consequences for an installation, plant and/or people.
These techniques assess, on the basis of numerical probability, each event in a chain of events. By aggregating these assessments, it is possible to arrive at an overall figure indicating the probability of human error for this complete chain of events.
The purpose of the assessment is to identify the measures necessary to reduce the likelihood of human error at certain points by introducing a range of changes, which may be of an organizational, procedural, ergonomic or other nature.

humidity The measure of the concentration of water vapour present in the atmosphere.

hydrant A permanent installation for providing water for fire fighting connected to a mains water supply. Access to the water main is by direct connection of a fire hose into a hydrant.

hygiene The science of health; rules for health.

hygrometer (anemometer) An instrument for measuring atmospheric humidity.

I

id In psychoanalytical theory, the aspect of personality concerned with instinctual reactions for satisfying motives. The id seeks immediate gratification of motives with little regard for the consequences or for the realities of life.

IEE Regulations (wiring regulations) A set of regulations produced by the Institution of Electrical Engineers aimed at ensuring compliance with the Electricity at Work Regulations 1989. Whilst these regulations have no legal status, they set safety standards for installers, designers, testers and erectors of electrical plant and equipment.

ignition temperature (ignition point) The temperature at which a small quantity of combustible material (fuel) will spontaneously ignite in a given atmosphere and continue to burn without any further input of heat.

illuminance The quantity of light flowing from a source, such as a light bulb. Sometimes referred to as 'luminous flux' or light flow, and measured in lux.

immunization The hardening of a person's attitude on a particular matter by giving him or her a mild exposure to an opposing attitude. This exposure hardens the originally held attitude so that it is resistant to change by the presentation of further facts or arguments, no matter how strong.

impact (shock) loading The sudden addition of a load to a material which produces a high rate of stress formation within the structure of the material. In certain cases, the stress will be dissipated to a surrounding structure. Where this dissipation does not occur, brittle fracture can result.

impedance The apparent resistance to current flow in an electrical circuit due to a magnetic effect (induction).

impinger A form of personal sampling instrument incorporating a glass bubble tube containing a liquid medium, which either reacts chemically to, or physically to dissolve, the contaminant. A known volume of air is passed through the impinger using a constant-flow sample pump and the liquid is analysed by gas chromatography or photospectrometry.

Improvement Notice An Improvement Notice may be served on an employer or other persons when, in the opinion of an enforcement officer, a person, business or undertaking is not complying with the relevant statutory provisions and where action is required by a certain date.
There is a right of appeal to an industrial tribunal against the notice, which suspends its enforcement until a decision is reached by the tribunal.

(Health and Safety at Work etc. Act 1974)
An Improvement Notice may be served where a fire authority is of the opinion that an occupier has not fulfilled his or her duty with regard to the provision of:

- means of escape in case of fire;
- means for fighting fire.

(Fire Safety and Safety of Places of Sport Act 1987)

impulse noise Noise which is produced by widely spaced impacts between, for instance, metal parts, such as drop hammers.

incident An event which does not result in injury, damage or loss, but which may cause interruption of the work process.
An undesired event that could, or does, result in loss.
An undesired event that could, or does, downgrade the efficiency of the business operation.

independent overriding control A safety control which cuts off the fuel supply to a gas-fired or oil-fired boiler, or air to the mechanical stokers of a solid fuel-fired boiler, when the water level in the boiler falls below a predetermined water level.

independent tied scaffold A scaffold featuring two rows of standards, where the total load is supported by the ground. It is tied to the building by a series of through ties to provide further stability.
Such a scaffold also incorporates transoms and diagonal bracing on the face and from the front row of standards to the rear row.

indication of danger This means, in relation to a substance or preparation dangerous for supply, one or more of the indications of danger referred to in Column 1 of Schedule 2 and:

(a) in the case of a substance dangerous for supply listed in Part I of the Approved Supply List, one or more indications of danger for that substance specified by a symbol-letter in Column 3 of Part V of that list; or
(b) in the case of a substance dangerous for supply not so listed, or a preparation dangerous for supply, one or more indications of danger determined in accordance with the classification of that substance or preparation under Regulation 5 and the Approved Classification and Labelling Guide.

[Chemicals (Hazard Information and Packaging for Supply) Regulations 1994]

indictable offences Where there is sufficient evidence, certain offences can, on the decision of a Magistrates' Court, be subject to committal proceedings through issue of an indictment, whereby an offender is committed to a Crown Court for trial.
Certain offences triable only on indictment are:

- breaching any of the relevant statutory provisions;

- acquiring, or attempting to acquire, possessing or using an explosive article or substance.

individual coaching A training technique entailing a one-to-one relationship between trainer and trainee. Coaching imparts knowledge, develops skills and forms attitudes during informal, but planned, sessions.

induction The logical process by which principles and rules are derived from observed facts.

induction training The Management of Health and Safety at Work Regulations 1999 place an absolute duty on employers to provide health and safety induction training for new employees.
Recommended topics for inclusion in induction training are:

- the organization's Statement of Health and Safety Policy;
- procedures for reporting hazards, accidents, near misses and ill health;
- details of hazards specific to the job, together with the precautions necessary;
- emergency procedures in the event of fire;
- current procedures for safety monitoring;
- welfare amenity provisions, e.g. sanitation, hand washing, showers, etc.;
- sources of health and safety information;
- role and function of health and safety specialists;
- correct use of personal protective equipment.

industrial tribunals Industrial tribunals deal with a wide range of industrial matters, including industrial relation issues, unfair dismissal, equal pay and sex discrimination.
Each tribunal consists of a legally qualified chairperson, appointed by the Lord Chancellor, and two lay members, one from management and one from a trade union.
Tribunals deal with the following employment/health and safety issues:

- appeals against Improvement and Prohibition Notices;
- time off for the training of safety representatives;
- failure by an employer to pay safety representatives for time off when undertaking their functions or for training;
- failure of an employer to make a medical suspension payment;
- dismissal, actual or constructive, following a breach of health and safety law and/or a term of an employment contract.

infiltration The entry of air into a building through openings in the fabric of the building.

informal consultation (1B5) Consultation between employer and employees which is unplanned and spontaneous, taking the form of discussion on a particular matter.

informal group A social group having no formal or permanent structure and consisting of people who happen to be assembled at a particular point in time.

information and advice sources Sources of health and safety information include:

- Acts of Parliament (Statutes);
- Regulations;
- HSC Approved Codes of Practice;
- HSE Guidance Notes;
- European Directives;
- HSC/HSE Consultative Documents;
- specific reports prepared by the HSE;
- British Standards, e.g. BS 5304 '*Safeguarding of Machinery*';
- safety data sheets supplied by manufacturers of hazardous substances;
- books, periodicals, films, videos, etc.;
- interviews and discussions;
- job descriptions;
- manufacturers' and suppliers' information and instructions;
- accident, illness and absence statistics.

Sources of professional advice include:

- HSE public enquiry points;
- local authority environmental health departments;
- public utilities;
- British Standards Institute;
- universities and colleges;
- national safety organizations, e.g. Royal Society for the Prevention of Accidents, National Radiological Protection Board;
- professional institutions, e.g. Institution of Occupational Safety and Health;
- consultants and consulting organizations.

information processing The relative speed with which people process information is a precondition of many accidents, or the errors or omissions which result in accidents.
Single Channel Theory explains the fact that the human mechanism for processing information has only finite capacity. Once a stimulus has been received by the brain, second and subsequent stimuli must wait until the first stimulus has been dealt with. Each stimulus produces a reaction or response.
Individual responses can be divided into two elements:

- specific reaction time, i.e. the actual time it takes to perceive and process the response;
- movement time, i.e. the time taken to move in order to execute the response.

Where a second stimulus arrives during the movement time of the first stimulus, that second stimulus must wait. Most people cannot do more than one thing

at a time, and the speed and sequence of response will vary from person to person.
With common tasks, such as operating a machine, typing on a keyboard or driving a car, the brain's monitoring action gradually reduces as familiarity with the task increases.

infrared A form of non-ionizing radiation emitted by all hot bodies. Long-term exposure may damage the eyes.

infrared analytical technique (2E1) An analytical technique which relies on the infrared absorption characteristics of certain chemical substances. Infrared radiation is passed through a cell which contains the contaminant under test. The absorption of radiation at specific wavelengths is proportional to the concentration of the contaminant.

ingestion The taking into the body of substances through the mouth.

inhalation The principal route of hazardous substances, in the form of gases, dusts, vapours, etc., into the body via the nose, mouth and respiratory tract.

initiation energy The energy necessary for a fire to start.

injection A forceful breach of the skin, perhaps as a result of injury, which can transmit hazardous substances past the skin barrier.

injury Damage to the body, commonly as a result of an accident.
Death or personal injury from electric shock, electric burns, electrical explosion or arcing, or from fire or explosion initiated by electrical energy, where any such death or injury is associated with the generation, provision, transmission, rectification, conversion, conduction, distribution, control, storage, measurement or use of electrical energy.
(Electricity at Work Regulations 1999)
Does not include injury caused by any toxic or corrosive substance that:

- has leaked or spilled from a load;
- is present on the surface of a load but has not leaked or spilled from it;
- is a constituent part of a load.

(Manual Handling Operations Regulations 1992)

'in motion or use' The phrase includes motion *or* use in the mode or manner in which the machine operates and for the purpose for which it is used, not merely movement of any kind, and does not include movement of the parts by hand in the case of a power-driven machine.
(Richard Thomas & Baldwin v. Cummings 919550 AC 321 & 1 AER 285)

in-running nip A particular machinery hazard associated with a trap formed, for instance, between a belt and a roller, drive wheel and chain, gear wheels and rollers. See Figure I.1.

inspection (work equipment) In relation to an inspection under Regulation 6:

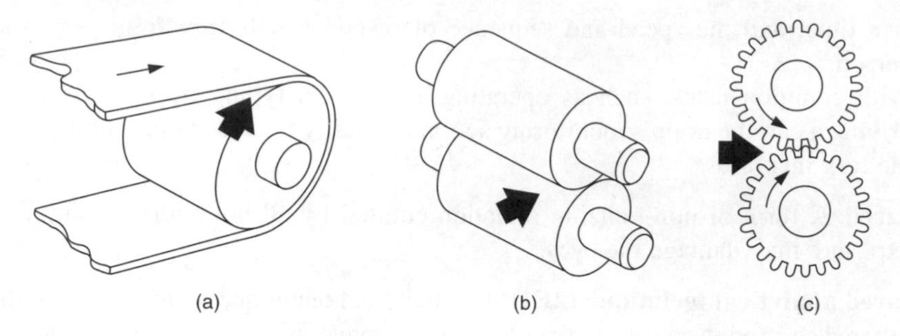

Figure I.1 In-running nips. (a) Between belt and pulley. (b) Between two rollers. (c) Between toothed wheels. Source: BS 5304 (1988) Safeguarding of Machinery.

- means such visual or more rigorous inspection by a competent person as is appropriate for the purpose described in that paragraph;
- where it is appropriate to carry out testing for the purpose, includes testing the nature and extent of which are appropriate for the purpose.

(Provision and Use of Work Equipment Regulations 1998)

insulation (2C2) Protection against direct contact with electrical conductors through the use of insulating materials with high resistance values (e.g. rubber, plastic), to prevent current flow through same.

insulation test In portable appliance testing, a test entailing application of a test voltage, typically 500 volts DC, between the live and neutral terminals bonded together and earth. From this test the insulation resistance is calculated by use of a portable appliance testing (PAT) device.

insulator A poor electrical conductor having a high resistance to the flow of electrical current, e.g. rubber, plastic.

insurance The Employers' Liability (Compulsory Insurance) Act 1969 requires employers to insure against claims by employees suffering personal injury, damage or loss.
The Certificate of Insurance must be displayed conspicuously at the workplace. The policy must state that any person under a contract of service or apprenticeship who sustains, injury, disease or death caused during the period of insurance and arising out of the course of employment will be covered for any legal liability on the part of the employer to pay compensation.

Integrated Pollution Control (IPC) A system of pollution control introduced in the Environmental Protection Act 1990. The main objectives of IPC are:

- to prevent or minimize the release of prescribed substances and to render harmless any such substances which are released;
- to develop an approach to pollution control that considers discharges from industrial processes to all media in the context of the effect on the environment as a whole.

intelligence A general term covering a person's abilities on a wide range of tasks involving mathematics, vocabulary, problem solving, etc. as measured by one of a variety of intelligence tests.

intention A significant element of a contact whereby the parties concerned must actually intend to enter into the legally binding agreement concerned.

interference A factor in learning and forgetting; the incompatibility of two learned associations.

intergranular attack A form of metal corrosion which occurs in alloys, where there is a combination of metals and/or carbon, and where the materials in the boundaries of the grains tend to be different from that in the centre. If the material at the boundary is liable to corrosion, the net effect is that a crack develops around the edges of the grains, leaving the grain boundaries porous.

interlocking guard A guard which has a movable part connected with the machinery controls so that:

- the part(s) of the machinery causing danger cannot be set in motion until the guard is closed;
- the power is switched off and the motion braked before the guard can be opened sufficiently to allow access to the dangerous part; and
- access to the danger point or area is denied while the danger exists.

(BS 5304: *Safeguarding of Machinery*)

International Commission for Radiological Protection (ICRP) (2E3) The ICRP is mainly concerned with setting dosage standards, covering all aspects of exposure from individual exposure to population exposure.
Maximum permissible concentrations are listed by the ICRP for air and water, i.e. air for breathing and drinking water.

International Loss Control Institute (ILCI) Loss Causation Model This model of accident causation takes into account all potential loss situations. Depicted as a series of five standing dominoes, the model indicates that:

1. Failure to maintain compliance with adequate standards (e.g. management training, organizational rules) can lead to
2. Deficiencies in personal factors (e.g. lack of knowledge) and job factors (e.g. inadequate purchasing), which can lead to
3. Substandard practices (e.g. failure to secure) and substandard conditions (e.g. noise exposure), which can lead to
4. Contacts, such as a fall to a lower level or being struck by something, which can lead to
5. Personal harm (e.g. major injury), property damage of varying degree and process loss of varying degree.

On this basis, the removal of one of the standing dominoes will result in the elimination or prevention of the items in domino number 5. See Figure I.2.

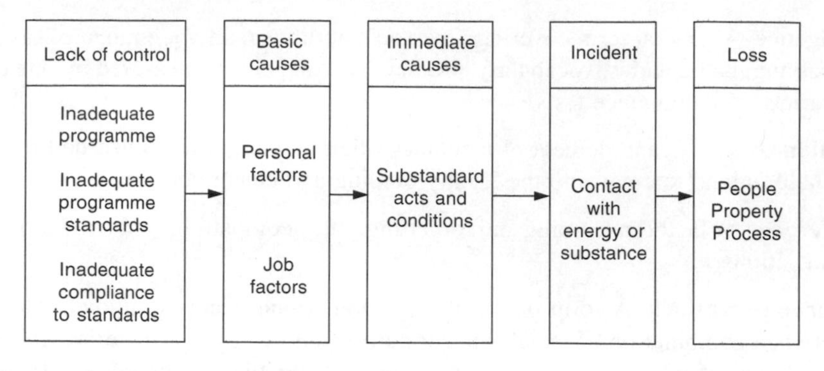

Figure I.2 ILCI Loss Causation Model.

intoxication The general state of harm caused by the effects on the body of a toxic substance.

intrinsically safe equipment (2C2) The use of electrical equipment in flammable atmospheres requires that such equipment should be intrinsically safe, i.e. not provide a source of ignition. On this basis, such equipment must be flameproofed and intrinsically safe for use in potentially flammable hazardous areas. These areas are classified according to a graded possibility of an explosive gas or vapour concentration occurring.

intrinsic safety (2C2) A concept based on the principle that sparks whose electrical parameters (voltage, current, energy) do not exceed certain levels are incapable of igniting a flammable atmosphere. It is applied as a concept to low-energy circuits, such as instrumentation and control systems.

ion A charged atom or group of atoms. It is charged by an excess or deficiency of electrons.

ionizing radiation (1E1) An ion is a charged atom or group of atoms. Ionization implies the charging of an atom or group of atoms to produce ions. Where the number of electrons does not equal the number of protons, the atom has a net positive or negative charge and is said to be ionized. Therefore, if a neutral atom loses an electron, a positively charged electron will result. Ionization is, therefore, the process of losing or gaining electrons and occurs in the course of many physical and chemical reactions.
Ionizing radiation includes alpha particles, beta particles, gamma rays, X-rays, neutrons, Bremsstrahlung and cosmic rays.

ionizing smoke detector (1C5) A form of smoke detector using a small radioactive source which maintains a level of ionization in two chambers, one of which is open to the atmosphere. When smoke enters this chamber, the smoke particles absorb some of the ionization, causing an electrical imbalance which is detected by the instrument.

irritant Non-corrosive substances and preparations which, through immediate

prolonged or repeated contact with the skin or mucous membrane, may cause inflammation.
[Chemicals (Hazard Information and Packaging for Supply) Regulations 1994]

isolation (1E3) A commonly used control measure against identified risks, implying the isolation or segregation of people from the particular hazard by, for instance, the use of remote control handling systems, enclosure of a plant or process producing harmful substances, the installation of high risk processing plants in remote parts of a country and enclosure of an individual in an acoustic booth or enclosure to protect against noise exposure.
The disconnection and separation of an electrical appliance from every source of electrical energy in such a way that both disconnection and separation are secure.

isotopes Atoms of the same element, having the same number of electrons and protons but having different numbers of neutrons. The isotopes of an element are identical in their chemical and physical properties, except those determined by the total mass of their atoms.

J

job enrichment A means of increasing satisfaction with, and the responsibilities of, a job, by reducing the degree of supervision or by allocating each individual a unit of work for which they have freedom to select the method and sequence of operations.

job impoverishment According to Herzberg, the removal of the individual interest, challenge and responsibility from a job.

Job Safety Analysis (1A3) A technique in the design of safe systems of work which identifies all the accident prevention measures appropriate to a particular job or area of work activity and the behavioural factors which most significantly influence whether or not these measures are taken.

job safety instructions (1A3) Job safety instructions are commonly one of the outcomes of job safety analysis, a technique used in the design of safe systems of work. Such instructions inform operators of specific risks at different stages of a job and advise of the precautions necessary to be taken at each stage.
Job safety instructions should be imparted to operators at the induction stage of their health and safety training and regularly reinforced.

Job Safety Review A means for reviewing compliance or otherwise with job safety instructions, with a view to determining whether there have been any deviations from the original instructions due to unforeseen factors, such as a change in materials or increased production requirements, and whether amendments to the instructions are required.

Job (task) Analysis The process of ascertaining what constitutes a particular job or task.

joint consultation An important means of improving motivation of people by enabling them to participate in planning work and setting objectives.
The process of consulting with employees and others on health and safety procedures and systems. This may take place through consultation by an employer with trade union-appointed safety representatives, non-trade union representatives of employee safety, and through the operation of a health and safety committee.
Legal and practical requirements relating to joint consultation are laid down in the Safety Representatives and Safety Committees Regulations 1977 and Health and Safety (Consultation with Employees) Regulations 1996, together with accompanying ACOP and HSE Guidance.

judicial precedent (1B1) Decisions made by judges, or precedents, have a certain

authority in the legal system. Most precedents are binding precedents, which means that the principle of law established in a former judicial decision will be binding in subsequent cases based on similar facts.
Precedents may be authoritative or persuasive.
Authoritative precedents are decisions which judges must follow.
Persuasive precedents are decisions which are not binding upon a court, but which a judge may take into consideration.

K

key exchange interlock (trapped key interlock) A form of interlock, using a master key, which ensures that the power to a machine is locked off when the guards are open or removed.

kier A fixed vessel used for boiling textile materials wherein boiling liquid is circulated by means of steam or mechanical power through a pipe, channel or duct, so constructed and arranged that the liquid is discharged over the textile material and percolates through it.

kinetics (1D3) The study of mechanical, nervous and psychological factors which influence the function and structure of the human body as a means of producing higher standards of skill and reducing cumulative strain.
A technique of moving in a more relaxed and efficient way.

Kinetic Theory of Gases When hydrogen reacts with oxygen, water is formed with the emission of heat. This reaction can be considered as molecules hitting one another and bouncing like balls or otherwise breaking up into different molecules. This 'ball' theory, known as the Kinetic Theory of Gases, proposes that the kinetic energy possessed by the balls, namely their speed, is a function of the temperature of the gas. The higher the temperature, the higher the kinetic energy of the molecules.

knowledge function One of the functions of attitude (Katz), implying people need to work within a particular framework, have a scale of values and generally know where they stand.

L

ladder scaffold A scaffold incorporating a working platform which is supported directly on a ladder or by means of a crutch or bracket on a rung or rungs of a ladder.

language Language is an important feature of the communication process. Language has a number of features:

- denotive – the key features that distinguish it;
- connotive – varies according to experience, association and context;
- indexical – provides an indication of the nature of the speaker.

laser A form of non-ionizing radiation consisting of very-high-energy light beams. The term 'laser' is an acronym derived from 'light amplification by stimulated emission of radiation'.

Law Reports A series of publications which record the judgements of the courts, e.g. the All England Reports (AER) and the Industrial Relations Law Reports (IRLR), and an important source of case law.

layout The arrangement of a particular workplace or work area, with particular reference to the space available for employees working in a particular room or area, and the situation of machinery, plant, equipment, processes and stored goods in relation to the employee and the work undertaken.

LC50 (2E5) The lethal concentration of a toxic substance in air which will kill 50 per cent of the test animal population of an experimental group (lethal concentration 50 per cent kill).

LD50 (2E5) The amount of toxic material which will kill 50 per cent of the test animal population of an experimental group (lethal dose 50 per cent kill).

learning A term referring to a relatively permanent change in behaviour that is the result of past experience or practice. It includes classical conditioning, operant conditioning and perceptual learning. See individual entries.

lecture A straight talk or exposition, possibly using visual or other aids, but without group participation other than through questions at the conclusion.

ledger In tubular steel scaffolding, a horizontal member tying a scaffold longitudinally.

legal liability (1A4) This term refers to the liability that one person may have to another (e.g. employer to employee, occupier to visitor or manufacturer of a product to a user of that product), under both criminal and civil law. See Figure L.1.

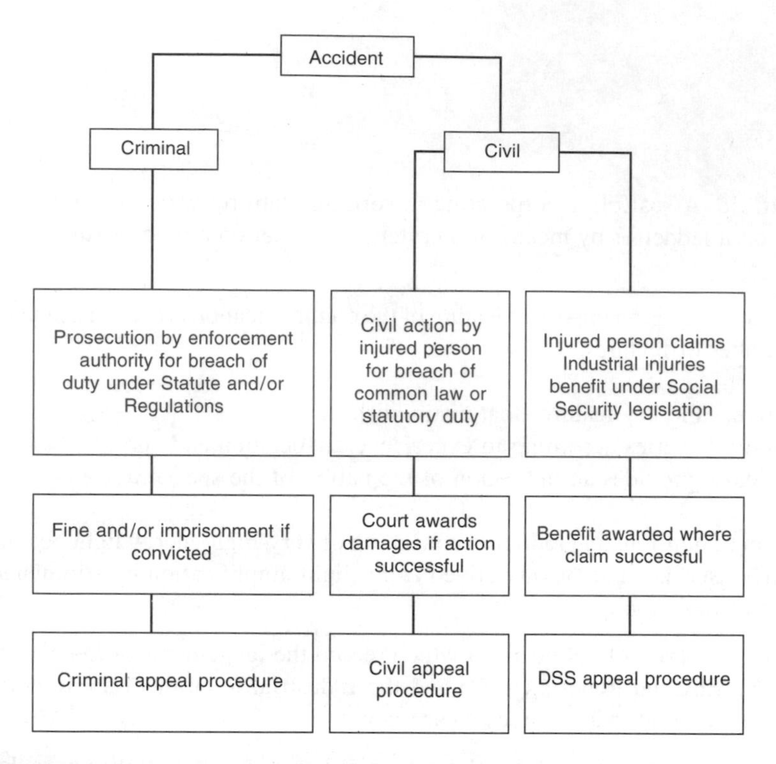

Figure L.1 Legal routes following an accident at work.

$L_{EP,d}$ and $L_{EP,w}$ (1E2) Terms used in the Schedule to the Noise at Work Regulations 1989 to indicate the daily and weekly average of daily personal noise exposure respectively of an employee. See Figure L.2.

L_{EQ} (equivalent continuous sound level) (1E2) Where sound pressure levels fluctuate, an equivalent sound pressure level, averaged over a normal 8-hour day.

lifting beam (spreader bar) (1D3) A specially designed steel beam incorporating a series of chains and hooks attached at different points, enabling difficult loads to be lifted in a particular way, often preventing horizontal stressing of eyebolts. Such a beam incorporates three parts, namely the beam, which is made from rolled steel section or plate, the means of attaching the beam to lifting machinery, such as a shackle, eyebolt or hole in the main structure of the beam, and the means of fixing the beam to the load, such as chain or wire rope slings.

lifting equipment (1D3) This means work equipment for lifting or lowering loads and includes its attachments used for anchoring, fixing or supporting it.
(Lifting Operations and Lifting Equipment Regulations 1998)
Lifting equipment includes blocks, tackles, gin wheels, jacks and winches.

THE SCHEDULE Regulations 2(1) and 13(1)

PART I

DAILY PERSONAL NOISE EXPOSURE OF EMPLOYEES

The daily personal noise exposure of an employee ($L_{EP,d}$) is expressed in dB(A) and is ascertained using the formula:

$$L_{EP,d} = 10 \log_{10} \frac{1}{T_0} \int_0^{T_e} \left[\frac{P_A(t)}{p_0}\right]^2 dt$$

where–

T_e = the duration of the person's personal exposure to sound;

T_0 = 8 hours = 28 800 seconds;

p_0 = 20 μPa; and

$P_A(t)$ = the time-varying value of A-weighted instantaneous sound pressure in pascals in the undisturbed field in air at atmospheric pressure to which the person is exposed (in the locations occupied during the day), or the pressure of the disturbed field adjacent to the person's head adjusted to provide a notional equivalent undisturbed field pressure.

PART II

WEEKLY AVERAGE OF DAILY PERSONAL NOISE EXPOSURE OF EMPLOYEES

The weekly average of an employee's daily personal noise exposure ($L_{EP,w}$) is expressed in dB(A) and is ascertained using the formula:

$$L_{EP,w} = 10 \log_{10} \left[\frac{1}{5} \sum_{k=1}^{k=m} 10^{0.1(L_{EP,d})_k}\right]$$

where–

$(L_{EP,d})_k$ = the values of $L_{EP,d}$ for each of the m working days in the week being considered.

Figure L.2 Formulae for assessing personal noise exposure.

lifting operation (1D3) This means an operation concerned with the lifting or lowering of a load.
(Lifting Operations and Lifting Equipment Regulations 1998)

lifting tackle This includes chain slings, rope slings, hooks, shackles and swivels.
(Factories Act 1961)

lift truck A form of mobile handling equipment which is operated either manually or driven using a form of power, such as electricity, compressed gas, petrol or diesel fuel.
Lift trucks may take a number of forms, depending upon the work to be undertaken, the loads carried and the working area in which they are operated.
Trucks are classified as pedestrian-operated stacking trucks, reach trucks, counterbalance trucks and narrow aisle trucks. See Figure L.3.

lighting (1C1) The flow of light from a source (illuminance), which may be natural or artificially produced, e.g. a window, light bulb or other forms of artificial lighting.

light obscuration detector (1C5) A form of smoke detector including a chamber incorporating a photo-electric cell opposite a light source, the cell detecting decreases in light intensity. When smoke enters the chamber, some of the light is obscured, actuating an alarm.

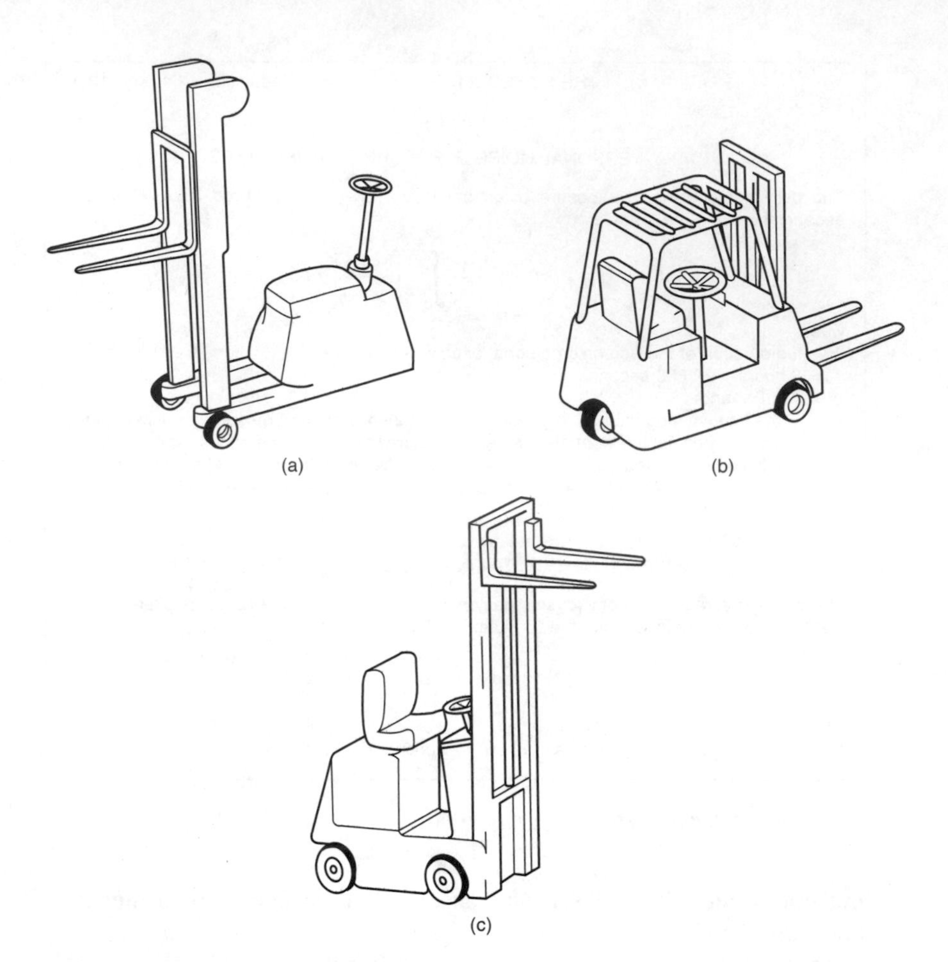

Figure L.3 Counterbalance fork trucks. (a) Lightweight pedestrian controlled. (b) Lightweight rider controlled. (c) Heavyweight rider controlled.

light scatter detector (1C5) A form of smoke detector incorporating a photo-electric cell fitted in a chamber at right angles to a source of light. When smoke enters the chamber, light is scattered and detected by the cell, actuating an alarm.

light work Work which does not jeopardize a child's safety, health, development, attendance at school or participation in work experience.
[Children (Protection at Work) Regulations 1998]

limitations of actions Civil actions may not be brought after the expiration of 6 years and, in the case of actions involving personal injury, 3 years.
In the case of injury arising from an accident at work, the limitation period is 3 years either from the date of the accident or from the deceased's personal representative's knowledge of the accident, whichever is the later.
Where the injured party dies within 3 years, the period recommences from the date of death or their personal representative's knowledge.
(Limitations Act 1980)

limits of flammability (2C1) There are two limits of flammability, the upper flammable limit and lower flammable limit. See individual entries.

liquefied flammable gas Any substance which would be a flammable gas at a temperature of 20°C and a pressure of 760 mm of mercury, but which is in a liquid form as a result of pressure or refrigeration, or both.
(Highly Flammable Liquids and Liquefied Petroleum Gases Regulations 1972)

liquefied fuels These include methane, butane, propane and hydrogen.

liquefied petroleum gas This covers both commercial butane and commercial propane, and any mixture of them.
(Highly Flammable Liquids and Liquefied Petroleum Gases Regulations 1972)

load Includes any person and any animal.
(Manual Handling Operations Regulations 1992)

load test In the testing of portable electrical appliances, a test which measures the load resistance between live and neutral terminals to ensure the resistance is not too low for safe operation of the appliance.

local effect An effect on the body of exposure to a toxic substance which is at the initial point of contact, e.g. the skin, nose, throat, bladder, eyes.

local exhaust ventilation (LEV) system Mechanical exhaust ventilation systems designed to intercept airborne contaminants at the point of, or close to, the source of generation, directing the contaminant into a system of ducting connected to an extraction fan and filtration unit.
LEV systems incorporate:

- a hood, enclosure or inlet to collect the agent;
- ductwork;
- a filter or air-cleaning device;
- a fan or other air-moving device;
- further ductwork to discharge clean air to the external air.

LEV systems may be of the receptor, captor or low-volume high-velocity type. See individual entries. See Figure L.4.

long-term memory A feature of memory concerned with the ability to store and subsequently recall information. It is developed from an early age through the repetition of items and codifying them to produce a meaning.

loss control Any intentional management action directed at the prevention, reduction or elimination of the pure (non-speculative) risks of business.
A management system designed to reduce or eliminate all aspects of accidental loss that lead to waste of an organization's assets.

Loss Control Management The application of professional management techniques and skills through those programme activities (directed at risk avoidance, loss prevention and loss reduction) specifically intended to minimize loss resulting from pure (non-speculative) risks of business.

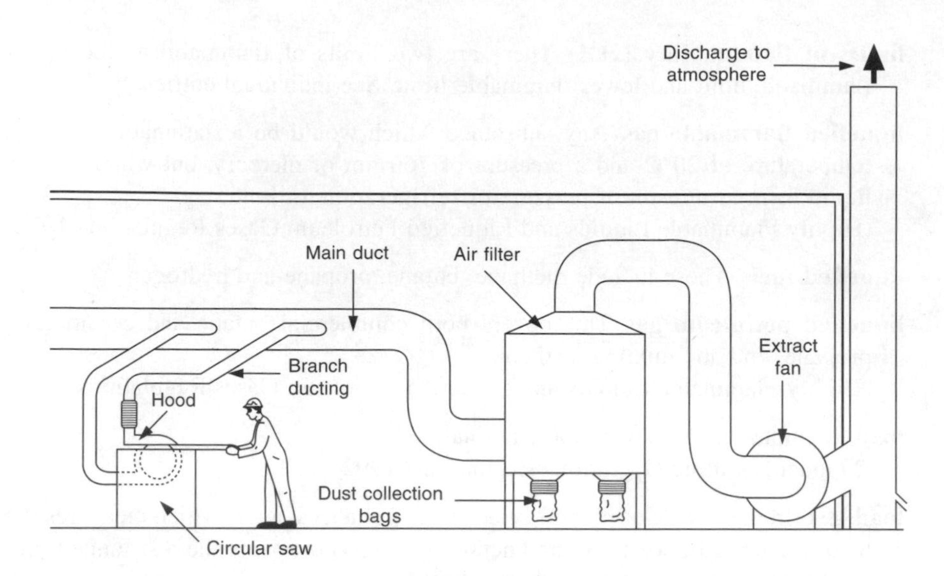

Figure L.4 A typical LEV system.

loss prevention The application of engineering techniques in order to reduce the occurrence of accidents that result in personal injury, damage to property, product, equipment and buildings, and those accidents that have no end result, i.e. near misses, but which downgrade the system.

loudness (2E2) A psychological attribute of tones, related to intensity but not directly proportional to it.

lower flammable limit (lower explosive limit) (2C1) The smallest concentration of flammable gas or vapour which, when mixed with air, is capable of ignition, and subsequent flame propagation, under prescribed test conditions.

low voltage This is a protective measure against electric shock. The most commonly reduced low voltage system is the 110-volt centre point earthed system. With this system the secondary winding of the transformer providing the 110-volt supply is centre tapped to earth, thus ensuring that at no part of the 100-volt circuit can the voltage to earth exceed 55 volts.

low-volume high-velocity (LVHV) system Certain types of hand-held tools such as high speed grinding and drilling machines, and pneumatically operated stone chipping equipment, produce large quantities of dust and require very high capture velocities at the point of dust production.
This is achieved by extraction through small apertures in the machine very close to the source of the contaminant. This generally entails the use of a cowl on the machine, located behind the point of production, to direct the dust towards the apertures. The dust passes through a hose to a collection bag.

Lux (1C1) The metric unit of luminous flux or illuminance, which equates to lumens per square metre.

lymphatic system An important protective system, the lymphatic system is a drainage system acting as a clearance channel for the removal of foreign bodies and bacteria from the body. Certain of these bacteria may be retained in the lymph nodes located at different points in the body, where a localized inflammation will arise.

M

machine (1D2) An apparatus for applying power, having fixed and moving parts, each with definite functions.
(BS 5304: *Safeguarding of Machinery*)

machinery hazards (1D2) A person may be injured at machinery through:

- coming into contact with, or being trapped between, the machinery and any material in or at the machinery or any fixed structure;
- being struck by, or becoming entangled in, motion in the machinery;
- being struck by parts of the machinery ejected from it;
- being struck by material ejected from the machinery (BS 5304).

The principal hazards associated with machinery are:

- traps – reciprocating and shearing traps, and in-running nips;
- entanglement – with unguarded rotating parts;
- ejection – of items from machines;
- contact – with, for instance, hot surfaces.

machinery guards (1D2) Safeguarding of machinery is achieved through a combination of physical guards and safety devices.
There are five main forms of machinery guard:

- A fixed guard, i.e. a guard which has no moving parts associated with, or dependent upon, the mechanism of any machinery, and which, when in position, prevents access to a danger point or area.
- An adjustable guard, i.e. a guard incorporating an adjustable element which, once adjusted, remains in that position during a particular operation.
- A distance guard, i.e. a guard which does not completely enclose a danger point or area but which places it out of normal reach.
- An interlocking guard, i.e. a guard which has a movable part so connected with the machinery controls that:
  - the parts of the machinery causing danger cannot be set in motion until the guard is closed;
  - the power is switched off and the motion braked before the guard can be opened sufficiently to allow access to the dangerous parts; and
  - access to the danger point or area is denied whilst the danger exists.
- An automatic guard, i.e. a guard which is associated with, and dependent

upon, the mechanism of the machinery and operates so as to remove physically from the danger area any part of a person exposed to the danger.

machinery reliability The reliability of machinery is directly related to the forces involved, such as differing loads and pressures, to which machines are subjected during normal service conditions, and the strength of materials as constructed or fabricated.

Machinery failures are commonly associated with failure of individual components or assemblies of components. Metal failure can result from a number of causes, such as ductile failure and creep. See individual entries.

machinery safety devices Safety devices take the form of:

- trip devices, i.e. a means whereby any approach by a person beyond the safe limit of working machinery causes the device to actuate and stop the machinery or reverse its motion, thus preventing or minimizing injury at the danger point;
- two-hand control devices, i.e. a device which requires both hands to operate the machinery controls, thus affording a measure of protection from danger only to the machinery operator and not other persons;
- overrun devices, i.e. a device which, used in conjunction with a guard, is designed to prevent access to machinery parts which are moving by their own inertia after the power supply has been interrupted so as to prevent danger;
- mechanical restraint devices, i.e. a device which applies mechanical restraint to a dangerous part of machinery which has been set in motion owing to failure of the machinery controls or other parts of the machinery, so as to prevent danger.

macrophages (phagocytes) Wandering scavenger cells, with an irregular outline and large nucleus, which move freely through tissue, engulfing bacteria and other particles. Macrophages secrete hydrolytic enzymes which attack the foreign bodies and bacteria.

magnetic inspection (2D4) A form of non-destructive testing whereby a component is magnetized and iron filings sprinkled on the surface; variations in the magnetic fields produced indicate flaws in the component.

maintainability A function of the rapidity and ease with which maintenance operations can be performed to help prevent malfunctions or to correct them if they occur.

maintenance (1C3) A combination of any actions carried out to retain an item in, or restore it to, an acceptable condition.

major accident An occurrence (including, in particular, a major emission, fire or explosion) resulting from uncontrolled developments in the course of the operation of any establishment and leading to serious danger to human health or the environment, immediate or delayed, inside or outside the establishment, and involving one or more dangerous substances.

(Control of Major Accident Hazards Regulations 1999)

major injury A major injury is classified as:

- Any fracture, other than to the fingers, thumbs or toes.
- Any amputation.
- Dislocation of the shoulder, hip, knee or spine.
- Loss of sight (whether temporary or permanent).
- A chemical or hot metal burn to the eye or any penetrating injury to the eye.
- Any injury resulting from electric shock or electrical burn (including any electrical burn caused by arcing or arcing products) leading to unconsciousness or requiring resuscitation or admittance to hospital for more than 24 hours.
- Any other injury:
  - leading to hypothermia, heat-induced illness or to unconsciousness;
  - requiring resuscitation; or
  - requiring admittance to hospital for more than 24 hours.
- Loss of consciousness caused by asphyxia or by exposure to a harmful substance or biological agent.
- Either of the following conditions which result from the absorption of any substance by inhalation, ingestion or through the skin:
  - acute illness requiring medical treatment;
  - loss of consciousness.
- Acute illness which requires medical treatment where there is reason to believe that this resulted from exposure to a biological agent or its toxins or infected material.

(Reporting of Injuries, Diseases and Dangerous Occurrences Regulations 1995, Schedule 1)

major spillage Certain major spillages are classed as dangerous occurrences under the Reporting of Injuries, Diseases and Dangerous Occurrences Regulations 1985. These are:

- the sudden uncontrolled release of 1 tonne or more of a highly flammable liquid from a plant;
- an uncontrolled release or escape of a specified dangerous substance from a road tanker or tank container.

malignant tumour A fast growing tumour which can infiltrate and extend into normal tissue and body structures, interfering with health and, in some cases, unless treated, causing death. They are spread by secondary deposits (metastases) to other sites in the body.

management The effective use of resources in the pursuit of organizational goals. 'Effective' implies achieving a balance between the risk of being in business and the cost of eliminating or reducing such risks.

Management Oversight and Risk Tree (MORT) (2A1) A systemic approach to the management of risks in an organization.

(United States Department of Energy)
A management technique aimed at increasing reliability, assessing the risks, controlling losses and allocating resources effectively.
The MORT philosophy is summarized thus:

1. Management takes risks of many kinds.
2. Risks in one area affect operations in other areas.
3. Risks should be made explicit where practicable.
4. Risk management tools should be flexible enough to suit a variety of adverse situations.

mandatory sign A sign which indicates that a specific course of action is required, e.g. EYE PROTECTION MUST BE WORN.
Mandatory signs are blue in colour, with a symbol or text in white, and circular in shape.

man hoist Within the meaning of section 25(2) of the Factories Act 1961, this is a continuous belt hoist for conveying persons between the floors of a building in both an upward and downward direction.
This form of hoist is commonly encountered in flour mills and other industrial premises where there is a need for employees to regularly move between floors to check various stages of a process.

manual handling injuries (1D3) More than a third of all industrial injuries result from manual handling operations.
HSE Guidance accompanying the Manual Handling Operations Regulations 1992 classifies the types of injury caused by manual handling accidents thus:

- sprains and strains 65%
- superficial injuries 9%
- contusions 7%
- lacerations 7%
- fractures 5%
- other forms of injury 5%.

Sites of bodily injury caused by manual handling are:

- backs 45%
- fingers and thumbs 16%
- arms 13%
- lower limbs 9%
- rest of torso 8%
- hands 6%
- other sites 3%.

What is significant, according to current statistical information, is that four out of five people will suffer some form of back condition at some time in their lives.

Principal manual handling injuries and conditions are muscle and ligamental strain, hernias, prolapsed (slipped) intervertebral discs and rheumatism.

manual handling operations (1D3) Any transporting or supporting of a load (including the lifting, putting down, pushing, pulling, carrying or moving thereof) by hand or bodily force.
(Manual Handling Operations Regulations 1992)

manual handling risk assessment (1D3) The Manual Handling Operations Regulations 1992 place a duty on employers to avoid, so far as is practicable, the need for employees to undertake manual handling operations at work which involve a risk of injury.
Where it is not reasonably practicable to avoid manual handling, an employer must make a suitable and sufficient risk assessment of the operations taking into account the following factors indicated in the Schedule to the regulations:

- the task;
- the load;
- the working environment;
- individual capability; and
- other factors, e.g. hindrance of movement by personal protective equipment;

and take appropriate steps to reduce the risk of injury to the lowest level reasonably practicable.
Employees must make full and proper use of any system of work provided by the employer in compliance with the steps taken by the employer. See Figure M.1.

maximum exposure limit (1E2) For a substance hazardous to health, this means the maximum exposure limit for that substance in relation to the specified reference period when calculated by a method approved by the HSC.
(Control of Substances Hazardous to Health Regulations 1999)
See ***Occupational Exposure Limits***.

means of escape (1C5, 2C1) A means of escape in case of fire is a continuous route by way of a space, room, corridor, staircase, doorway or other means of passage, along or through which persons can travel from wherever they are in a building to the safety of the open air at ground level by their own unaided efforts.

measurement technique (1A5) A technique for measuring safety performance by means of safety audits, inspections, surveys, safety tours and safety sampling exercises. See individual entries.

mechanical interlocking A safety device for machinery incorporating two specific elements: firstly, the operation or actuation of a device, which may be a hydraulic or pneumatic valve; and, secondly, as a result of the operation of the device, the movement of a particular component, generally a guard.

mechanical restraint device A device which applies mechanical restraint to a dangerous part of machinery which has been set in motion owing to failure of the machinery controls or other parts of the machinery, so as to prevent danger.

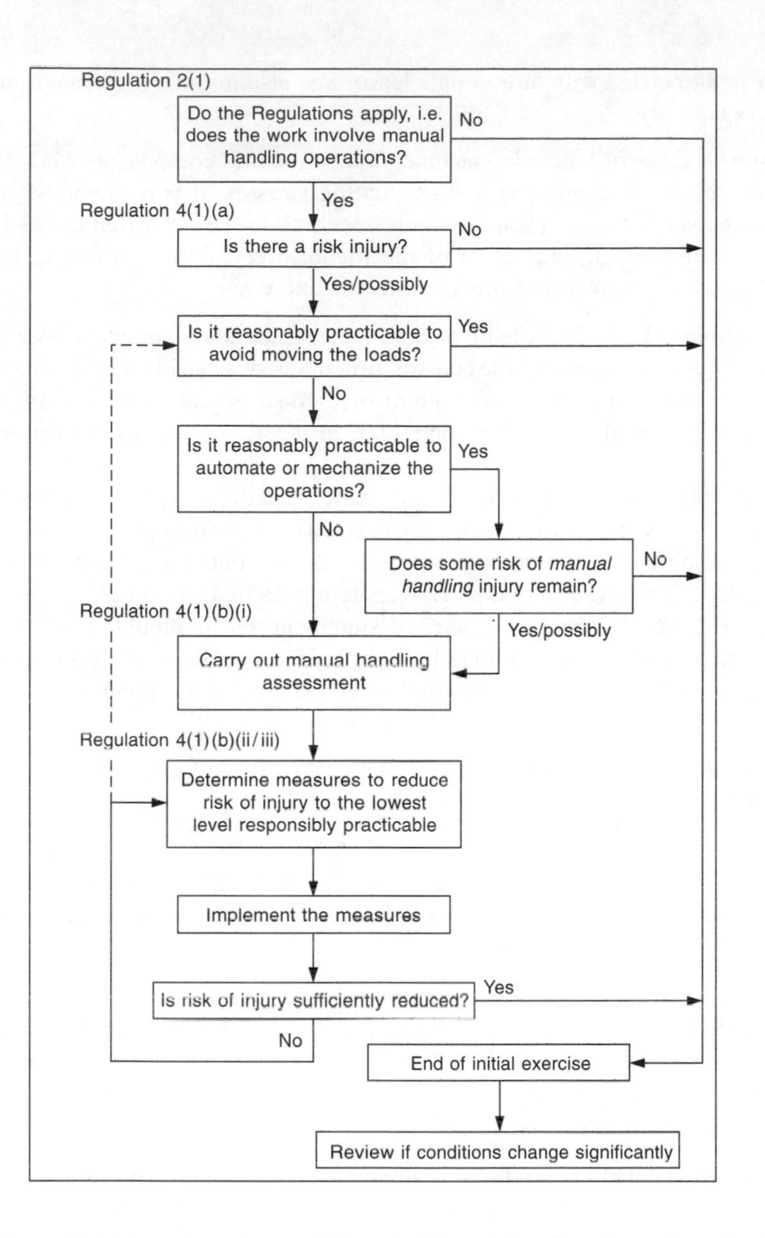

Figure M.1 Flow chart: how to follow the Manual Handling Operations Regulations 1992. Source: Health and Safety Executive.

(BS 5304: *Safeguarding of Machinery*)

mechanical trip device A safety device incorporating a barrier which is contacted by a part of the body as it approaches the danger area of a machine. Contact with the barrier operates this device which brings the machine to rest.

memory The process of retaining, recognizing and recalling experience. It is particularly associated with how people learn. See also ***long-term memory*** and ***short-term memory***.

metal fatigue A cause of failure in machinery and structural components whereby metals are subject to continuing and fluctuating stresses. It is common where metals are subject to mechanical vibration, resulting in the formation of cracks which can penetrate the full extent of the metal. Metal fatigue is one of the primary failure mechanisms of pressure vessels and cranes.

method statement (2C4) The use of method statements is a common feature of safety practices in, particularly, the construction industry. Considerable emphasis is placed on the use and implementation of method statements in procedures to ensure compliance with the Construction (Design and Management) Regulations 1994.

A method statement is a formally written safe system of work, or series of integrating safe systems of work, agreed between a client or occupier of premises and a contractor, or between a main contractor or subcontractor, and produced where work with a foreseeably high risk content is to be carried out.

There is no standard format for a method statement, but it should specify the operations to be undertaken on a stage-by-stage basis and indicate the precautions necessary to protect site operators, employees occupying the premises where work is undertaken, and members of the public.

micro-organism A microbiological entity, cellular or non-cellular, which is capable of replication or transferring genetic material.
(Control of Substances Hazardous to Health Regulations 1999)

microwaves A form of non-ionizing radiation emitted at very high radio frequencies. Microwaves can penetrate the body, causing heating of internal organs, in particular the eyes and testes.

mine An excavation or system of excavations made for the purpose of, or in connection with, the getting, wholly or substantially by means involving the employment of persons below ground, of minerals (whether in the natural state or in solution or suspension) or products of minerals.
(Mines and Quarries Act 1954)

minimum ignition energy The minimum amount of energy, measured in millijoules (mJ), necessary for ignition of a particular substance, e.g. hydrogen – 0.019 mJ, methane – 0.39 mJ.

minimum measured illuminance The minimum measure of illuminance acceptable for a particular type of work.

misconduct This is a term implying some form of behaviour which is deemed unacceptable and which could result in some form of disciplinary action against an employee by the employer. There is a difference between 'misconduct' and 'gross misconduct'.

Misconduct is action which is deemed a breach of the workplace rules but not sufficiently serious to merit instant or summary dismissal, but which would result in a formal warning, e.g. non-observance of safety procedures, persistent lateness.

Gross misconduct is an act which is so serious as to amount to the employee 'smashing the employment contract', with the result that dismissal without notice is deemed to be appropriate in the circumstances, e.g. physical assault, theft, fraud.

Company rules should specify aspects of behaviour which constitute misconduct and gross misconduct.

mist A finely dispersed liquid created by processes such as spraying, foaming, pickling and electroplating.

mobile access equipment (towers) Used particularly in construction and maintenance work, this equipment may take the form of a movable tower either formed from scaffold tubes or constructed from pre-formed frames which interlock one above the other. In each case, the tower incorporates a working platform, access by means of internal fixed vertical or raking ladders, together with castors for ease of movement of the tower.

To ensure safe working:

- the working platform should be fitted with guard rails and toe boards;
- the platform should be secure and completely boarded;
- the height of the working platform must not exceed three times the smaller base dimension; where this dimension is exceeded, outriggers should be fitted;
- outriggers should be installed when working in windy conditions;
- the rigidity of the tower should be increased by diagonal bracing;
- castors should be fitted with effective brake devices;
- towers should be moved with great care; no one should be allowed on the working platform during movement of the tower;
- the tower should be moved by pushing or pulling at base level.

mobile elevating platform (1C3) An elevated working platform, mounted on a vehicle chassis, which can be driven from one location to another. Such platforms operate on a scissor-lift and telescopic mode of action, and are commonly used for street lighting maintenance and other forms of high level maintenance.

modelling A behaviour modification technique which depends upon imitation and perceptual learning.

Modelling Approach A theory of accident causation developed by Hale and Hale (1970), who developed a simulation of the accident process.

In Hale and Hale's model, use is made of a closed loop system, which considers the major factors of presented, expected and perceived information, possible actions, the decision-making process, the final action and the feedback loop. See Figure M.2.

motivation (1A2, 2A2) Behaviour instigated by needs and directed towards goals. A motivator is something that produces the drive to produce certain behaviour or to mould behaviour.

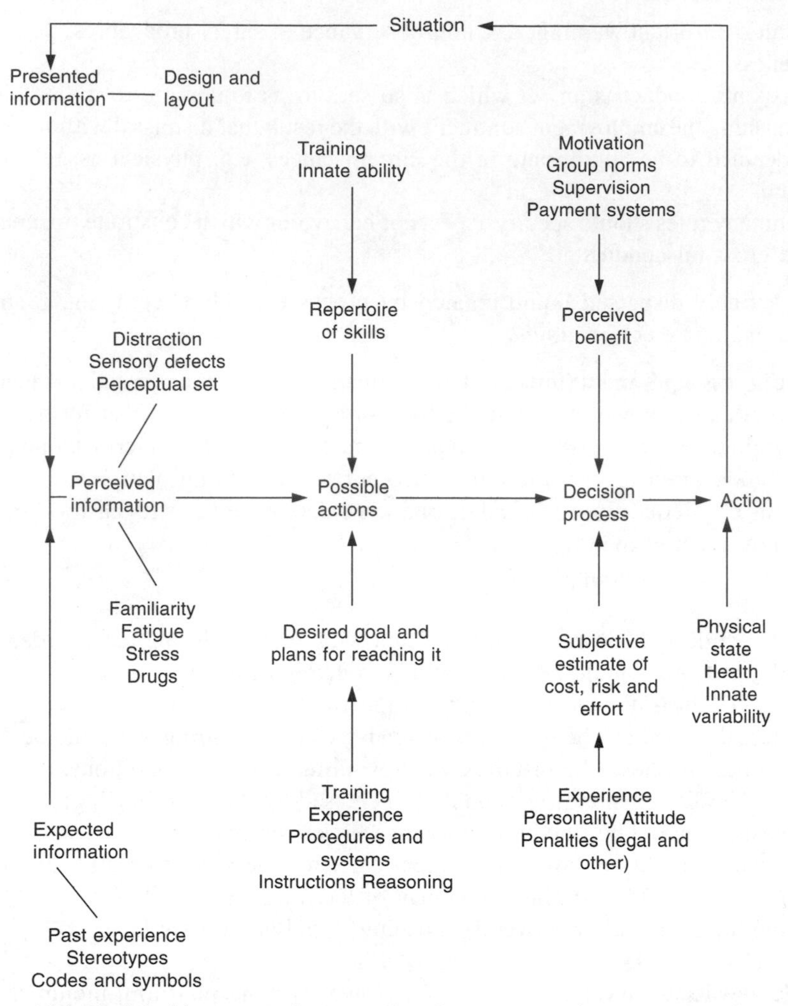

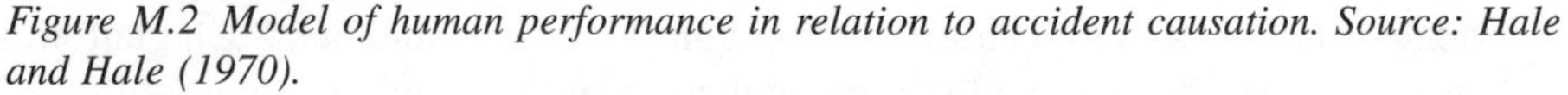

Figure M.2 Model of human performance in relation to accident causation. Source: Hale and Hale (1970).

'Motivation' describes the goals, aims or objectives that people endeavour to meet and the drive or motivating force that keeps them on track in pursuing these goals or objectives.

motive A need and the direction of behaviour towards a particular goal.

moving walkways (travelators) A form of continuous slow moving horizontal belt for conveying people, commonly found in airports.
Moving walkways shall:

- function safely;
- be equipped with any necessary safety devices;
- be fitted with one or more emergency stop controls which are easily identifiable and readily accessible.

[Workplace (Health, Safety and Welfare) Regulations 1992]

multi-load A load consisting of two or more dangerous substances in separate compartments or tanks.

Multiple Causality (Causation) Theory Multiple causality means that there may be more than one cause of an accident.

This theory of the causes of accidents takes into account two principal features of the events leading to an accident, namely an unsafe act and an unsafe condition. However, there are a number of causes (multiple causes) which contribute or lead to both the unsafe act and the unsafe condition. These contributory causes come together in random fashion, resulting in the accident.

By identifying these multiple causes, the unsafe act or unsafe condition should be prevented from arising. See Figure M.3.

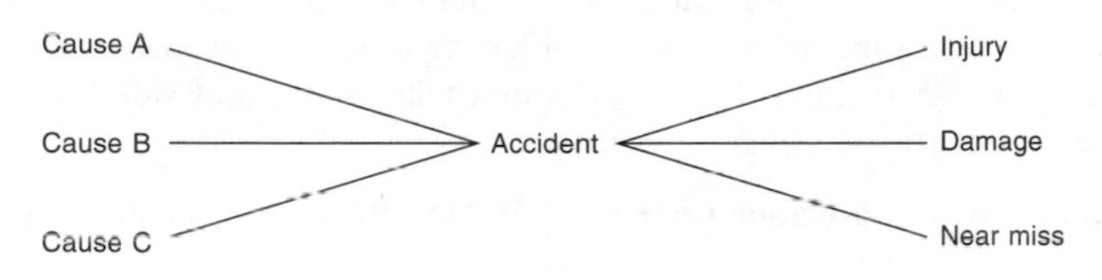

Figure M.3 Multiple causality theory.

mutagenesis (2E1) The alteration of the genetic material of a cell in such a manner that the alteration is transmitted to subsequent generations of cells.

mutagenic Substances and preparations which, if they are inhaled or ingested or if they penetrate the skin, may induce heritable genetic defects or increase their incidence.

[Chemicals (Hazard Information and Packaging for Supply) Regulations 1994]

mutagenicity (2E5) The ability of a substance or preparation to induce heritable genetic defects or increase their incidence.

mutation A change in a gene and in the characteristic it determines.

N

narcosis A condition in which the effects of toxic substance exposure cause a reduction or dulling in consciousness, an anaesthetic effect.

narcotic substance An organic substance which may be absorbed into the bloodstream, producing an anaesthetic effect.

narrow aisle truck A type of lift truck used in storage depots and warehouses where the truck moves in one aisle only, not having to turn. This enables close positioning of storage racking. These trucks may take the form of side loaders and counterbalance rotating load turret trucks.

National Radiological Protection Board (2E3) This Board has the following functions:

- by means of research and otherwise, to advance the acquisition of knowledge about the protection of mankind from radiation hazards;
- to provide information and advice to persons, including Government Departments, with responsibilities in the UK in relation to the protection from radiation hazards either of the community as a whole or particular sections of the community.

The Board maintains a register of all radiation workers, except those employed by the Ministry of Defence. In fulfilment of its statutory function, the Board produces reports from time to time reviewing the radiation exposure of the population.

near miss An unplanned and unforeseeable event that could have resulted in death, human injury, property damage or other form of loss.

necking (2D4) The reduction in cross-sectional area at a particular point in a material where the load applied has exceeded the ultimate tensile strength of the material.
Necked stays are commonly associated with boiler failure.

necrosis Death of tissue, destruction of cells.

negative reinforcement A form of reinforcement occurring when the learner escapes from, or avoids, an unpleasant stimulus.

negligence (1B1, 2B3) 'Negligence' is commonly defined as 'careless conduct injuring another'.
It is defined in common law as:

- the existence of a duty of care owed by the defendant to the plaintiff;
- breach of that duty; and
- injury, damage or loss resulting from or caused by that breach.

(Lochgelly Iron & Coal Co. Ltd. v. M'Mullan, 1934)
All the above criteria must be established by a claimant before a civil claim can be brought.

'neighbour principle' (1B1, 2B3) This principle was established in Commissioner for Railways v. McDermott (1967) 1 AC 169. Lord Gardner explained the position with regard to occupation of premises thus:
'Occupation of premises is a ground of liability and is not a ground of exemption of liability. It is a ground of liability because it gives some control over and knowledge of the state of the premises, and it is natural and right that the occupier should have some degree of responsibility for the safety of persons entering his premises with his permission. There is proximity between the occupier and such persons and they are his neighbours. Thus arises a duty of care.'

nervous system (1E1) The brain, spinal cord and nerves serving the various sense organs, muscles and endocrine glands.

neutralization A control strategy for hazardous substances whereby a neutralizing compound is added to a highly dangerous compound, e.g. acid to alkali, thereby reducing the immediate danger. Many hazardous wastes are neutralized prior to transportation.

neutrons A form of ionizing radiation, a neutron is an elementary particle with unit mass and no electric charge.

new or expectant mother An employee who is pregnant, who has given birth within the previous 6 months, or who is breastfeeding.
(Management of Health and Safety at Work Regulations 1999)

'no case to answer' A defence submission in a Crown Court following the reading of the prosecution's charge. Where this submission is accepted by the judge, the jury are directed to find the defendant 'not guilty'.

no fault liability (2B3) This term implies that there is no requirement to establish fault or intent in that a claimant is automatically compensated for injury, damage or loss by the state or some other organization. It thus dispenses with or disregards the common law and statutory elements of liability.

noise 'Noise' is variously defined as 'unwanted sound' or 'a loud sound'.
Noise can:

- be a nuisance, resulting in disturbance and loss of enjoyment of life;
- be a contributory factor in accidents, e.g. masking out warning signals; and
- cause hearing impairment.

The most common condition associated with exposure to noise is occupational deafness.

noise exposure record A formal record of the exposure of persons at work to noise, which includes the following details:

- address of premises;
- date of noise survey;
- name of person undertaking the survey;
- number of persons exposed in the workplace;
- noise level (leq or sound pressure level);
- daily exposure period(s)
- $L_{EP,d}$ – dB(A)
- peak pressure (where appropriate);
- comments/remarks of assessor;
- instruments used, including date of last calibration;
- the signature of the assessor and date of the record.

(Noise at Work Regulations 1989)
See Figure N.1.

Name and address of premises, department etc.____________________

Date of survey__________ Survey made by____________________

Workplace Number of persons exposed	Noise level (Leq(s) or sound level)	Daily exposure period	$L_{EP,d}$ dB(A)	Peak pressure (where appropriate)	Comments/ remarks

General comments ____________________

Instruments used ____________________

Date of last calibration__________ Signature__________

Date

Figure N.1 Record of noise exposure. Source: Noise at Work, Noise Guide No. 3: Noise Assessment, Information and Control. London: HMSO.

noise immission The total sound energy impinging on the ear.

noise-induced hearing loss (occupational deafness) Exposure to noise at work may affect hearing in three ways:

- temporary threshold shift: a short-term effect, i.e. a temporary reduction in the ability to hear, which may follow exposure to excessive noise, such as that from rifle fire or certain types of machinery, such as chain saws;
- permanent threshold shift: a permanent effect where the limit of tolerance is exceeded in terms of the duration and level of exposure to noise and individual susceptibility to noise;
- acoustic trauma: a condition which involves sudden damage to the ear from short-term intense exposure or even from one single exposure, e.g. gun fire, major explosions.

Noise-induced hearing loss is a prescribed occupational disease.

non-destructive testing (2D4) Non-destructive testing techniques supplement destructive testing techniques, such as tensile and compressive tests. The purposes of non-destructive tests are:

- to establish the proportions, dimensions and fitness for manufacture of components;
- for checking the durability or deterioration of structures.

A number of techniques are used for the inspection of surfaces of machinery, equipment and components:

- visual examination;
- optical examination using a small binocular microscope;
- mechanical examination;
- ultrasonic techniques, which entail transmitting waves of certain frequencies to observe reflections which indicate flaws and defects;
- magnetic techniques, whereby the component is magnetized and iron filings sprinkled on the surface; variations in the magnetic field produced indicate flaws;
- electrical resistivity tests, based on the time taken for a current to pass around a flaw or crack being greater than where no flaw or crack existed;
- the use of dye penetrants, whereby a crack is coated with oil containing a dye, followed by visual observation to ascertain the oil entering and leaving the crack under examination.

non-ionizing radiation (1E1, 2E3) This comprises electromagnetic radiation at the longer end of the electromagnetic spectrum.

non-mechanical hazards (1D2) Those hazards associated with machinery but not arising from machinery motion, e.g. risk of burns from hot surfaces, contact with hazardous substances used in machines, exposure to machinery noise and airborne contaminants emitted from machines.

non-operational parts (machinery) Those functional parts of machinery which convey power or motion to the operational parts, e.g. transmission machinery.

non-verbal communication (1CS1) That form of communication which may be of a written, visual, tactile or auditory nature.
Non-verbal communication can be used to support verbal communication and may take the form of gestures, glances, terminal glances, facial expression, a particular posture and involuntary features, such as blushing, pallor, perspiration and tension.

notice of intent A preliminary form of notice served by an enforcement authority advising the recipient of the intention of that enforcement authority to serve an enforcement notice where there is a failure to comply with any provision of the workplace fire precautions legislation.
[Fire Precautions (Workplace) Regulations 1997]

nuisance at common law A nuisance which, apart from statute, violates the principles which the common law lays down for the protection of the public and of individuals in the exercise and enjoyment of their rights.

nuisance (statutory) The following are statutory nuisances:

- any premises that are in such a state as to be prejudicial to health or cause a nuisance;
- smoke emitted from premises that is prejudicial to health or causes a nuisance;
- fumes or gases emitted from premises that are prejudicial to health or cause a nuisance;
- any dust, steam, smell or other effluent arising on industrial, trade or business premises that are prejudicial to health or cause a nuisance;
- any accumulation or deposit that is prejudicial to health or causes a nuisance;
- any animal kept in a place or manner that is prejudicial to health or a nuisance;
- noise emitted from premises that is prejudicial to health or causes a nuisance;
- any other matter that is declared by any other enactment to be a statutory nuisance.

(Environmental Protection Act 1990)

O

obiter dicta A feature of a judgement implying 'comments by the way', such as a reference to what could have happened had the circumstances been different. *Obiter dicta* are 'persuasive' but not 'binding'. In certain cases, the *obiter dicta* can be so persuasive as to be incorporated in subsequent or later judgement and become part of the *ratio decidendi.*

occupational deafness See ***noise-induced hearing loss***.

Occupational Exposure Limits (1E2) HSE Guidance Note EH 40 'Occupational Exposure Limits', which is regularly revised, gives details of Occupational Exposure Limits (OELs) which should be used for the purposes of determining the adequacy of control of exposure by inhalation to substances hazardous to health. These limits form part of the requirements of the Control of Substances Hazardous to Health (COSHH) Regulations 1999.

OELs are classified as either Maximum Exposure Limits (MELs) or Occupational Exposure Standards (OESs).

Both MELs and OESs may have stated Short-Term Exposure Limits (STELs) and Long-Term Exposure Limits (LTELs).

MELs are listed in both Schedule 1 of the COSHH Regulations and Table 1 of Guidance Note EH 40.

An MEL is the maximum concentration of an airborne substance, averaged over a reference period, to which employees may be exposed by inhalation under any circumstances and is specified, together with the appropriate reference period, in Schedule 1 of the COSHH Regulations.

An OES is the concentration of an airborne substance, averaged over a reference period, at which, according to current knowledge, there is no evidence that it is likely to be injurious to employees if they are exposed by inhalation, day after day, to that concentration and which is specified in a list approved by the HSC.

The STEL is aimed primarily at avoiding acute effects, or at least reducing the risk of the occurrence, and relates to a 10-minute reference period in any 24 hours.

The LTEL is concerned with the total intake over long periods, normally an 8-hour reference period, and is therefore appropriate for protecting against the effects of long-term exposure.

Certain substances listed in Guidance Note EH 40 carry a 'Skin' annotation (Sk), implying that the substance can be absorbed through the skin.

Concentrations of gases and vapours in air are usually expressed as parts per million (ppm), a measure of concentration by volume, as well as in milligrams per cubic metre, a measure of concentration by mass. Concentrations of airborne particles, such as fumes and dusts, are usually expressed in milligrams per cubic metre. Concentrations of mineral fibres are expressed as fibres per millilitre of air.

Occupational Exposure Standard (1E2) See ***Occupational Exposure Limits***.

occupational health (1E1) This is variously defined as:

- a branch of preventive medicine concerned with health problems caused by or manifest at work;
- a branch of preventive medicine concerned with the relationship of work to health and the effects of work upon the worker.

occupational health nurse (1E3) Generally taken to imply a Registered General Nurse (RGN) with a recognized qualification in occupational health nursing, such as the Occupational Health Nursing Certificate of the Royal College of Nursing.
The occupational health nurse's role consists of eight main elements:

- health supervision;
- health education;
- environmental monitoring and occupational safety;
- counselling;
- treatment services;
- rehabilitation and resettlement;
- unit administration and record systems;
- liaison with other agencies, e.g. the EMAS.

occupational health physician (1E3) A registered general medical practitioner, preferably with a postgraduate qualification in occupational medicine.

occupational hygiene (1E1) This is defined as the identification, measurement and control of contaminants and other phenomena, such as noise and radiation, which would otherwise have unacceptable adverse effects on the health of people exposed to them.
The four principal areas of occupational hygiene practice are:

- identification/recognition of the specific contaminant;
- measurement, using an appropriate measuring technique;
- evaluation against an existing standard, e.g. Occupational Exposure Limits;
- prevention or control of exposure.

occupational hygienist (1E3) A specialist principally concerned with the practical measurement, monitoring, evaluation and control of health risks to employees and others, e.g. exposure to noise, radiation, toxic substances and other forms of environmental hazards.
Entry to the profession is controlled by the British Examination and Registration Board in Occupational Hygiene (BERBOH).

occupiers' liability (1B1) This is concerned with the duties of people and organizations who occupy land and premises, whose land and premises are visited by people for a variety of purposes.

Section 4 of the HSWA requires that people in control of premises take reasonable care towards others who may enter their premises.

Anyone who is injured while visiting or working on land or premises may be in a position to sue the occupier for damages.

Occupiers' liability is dealt with in the Occupiers' Liability Acts 1957 and 1984 (OLA).

The OLA 1957 established a common duty of care to all lawful visitors, namely a duty to take such care as in all the circumstances of the case is reasonable to see that the visitor will be reasonably safe in using the premises for the purposes for which he or she is invited or permitted by the occupier to be there.

Section 1 of the OLA 1957 defines the duty of occupiers to all persons lawfully on the premises in respect of dangers due to the state of the premises or to things done or omitted to be done on them.

Visitors to premises are classed as both invitees and licensees. Protection is afforded to all lawful visitors, whether they enter for the occupier's benefit, such as customers, or for their own benefit, such as an enforcement officer, though not to persons exercising a public or private right of way over premises.

Under the OLA 1957, occupiers have a duty to erect warning notices, but a warning notice does not in itself absolve the occupier from liability unless in all the circumstances it is sufficient to enable the visitor to be reasonably safe.

An occupier must be prepared for children to be less careful than adults and must be aware of any lure or attraction to children, such as a pond, that could constitute a trap.

The OLA 1984 imposes a duty on the occupier in respect of trespassers, namely persons who may have lawful authority to be in the vicinity or not, who may be at risk of injury on the occupier's premises. The above duty can be discharged by issuing some form of warning, such as the display of a notice, but such a notice must be very explicit and its requirements actively enforced by the occupier.

Octave Band Analysis A sound measurement technique which enables the way sound is distributed throughout the frequency spectrum to be identified. The sound is divided into octave bands and measured at the geometric centre frequency of each band.

Octave band analysis is used for assessing the risk of occupational deafness, in the analysis of machinery noise and specification of remedial measures, and in the specification of certain types of hearing protection.

offences under the law (1B2) Offences arising from a breach of the Health and Safety at Work etc. Act 1974 and any of the relevant statutory provisions.

Ohm's law At a constant temperature the current through a conductor is directly proportional to the potential difference across its ends.

Thus:

$$\text{Resistance (Ohms)} = \frac{\text{Pressure}}{\text{Current}} = \frac{\text{Volts}}{\text{Amps}}$$

Volts = Amps × Ohms

operant conditioning Learning to make a particular response to secure positive reinforcement or to escape or avoid negative reinforcement.

operator A self-employed person who habitually uses display screen equipment as a significant part of his normal work.
[Health and Safety (Display Screen Equipment) Regulations 1992]

Orange Book This is the United Nations publication 'Recommendations on the Transport of Dangerous Goods', which provides guidance on a range of issues affecting such transport operations.
The Orange Book provides a system for the classification of chemical substances which are considered to be dangerous when transported. Each substance is given a four figure number. All substances listed are classified according to a particular hazard class, of which there are nine classes, together with a diamond-shaped symbol.

order pickers (1D3) A device originally derived from the fork lift truck, incorporating a working platform permanently fixed to the tines of a truck. The truck is operated from the picking platform or, in some cases, the truck moves along a fixed traverse only. See Figure O.1.

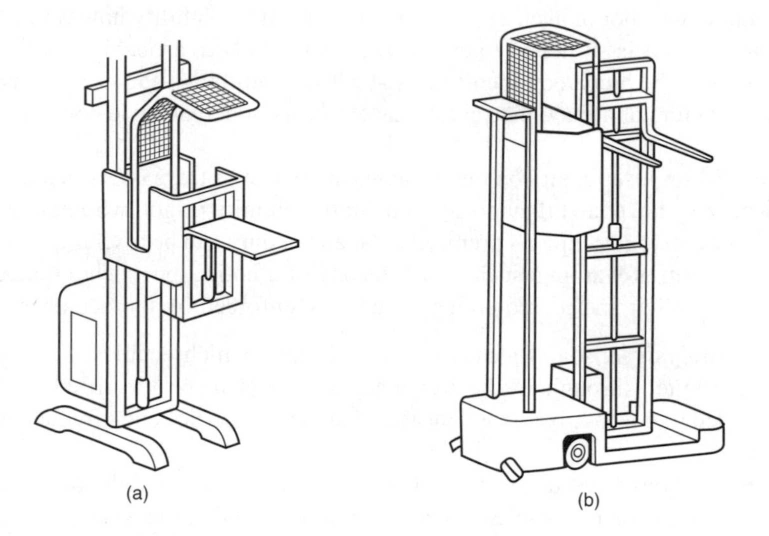

Figure O.1 Order pickers. (a) Conventional. (b) Purpose designed.

organism Any acellular, unicellular or multicellular entity (in any form), other than humans or human embryos, and, unless the context otherwise requires, the term also includes any article or substance consisting of biological matter.
(Environmental Protection Act 1990)

orifice plate A device for the measurement of air velocity in a duct. Air velocity can be calculated from the pressure drop measured across the orifice plate which is fixed in the duct.

overrun device A device which, used in conjunction with a guard, is designed to prevent access to machinery parts which are moving by their own inertia after the power supply has been interrupted, so as to prevent danger.
(BS 5304: *Safeguarding of Machinery*)

oxidation The exposure of a metal surface to oxygen and the subsequent chemical reaction which takes place, particularly at high temperatures. Oxidation causes deterioration of the surface and eventual breaking up of the surface.

oxidizing Substances and preparations which give rise to an exothermic reaction in contact with other substances, particularly flammable substances.
[Chemicals (Hazard Information and Packaging for Supply) Regulations 1994]

oxygen monitor A device which monitors the percentage oxygen present in the atmosphere at any point in time, commonly linked to a visual and/or audible alarm.

P

package In relation to a substance or preparation dangerous for supply, the package in which the substance or preparation is supplied, and which is liable to be individually handled during the course of the supply and includes the receptacle containing the substance or preparation and any other packaging associated with it and any pallet or other device which enables more than one receptacle containing a substance or preparation dangerous for supply to be handled as a unit, but does not include:

- a freight container (other than a tank container), a skip, a vehicle or other article of transport equipment; or
- in the case of supply by way of retail sale, any wrapping such as a paper or plastic bag into which the package is placed when it is presented to the purchases.

[Chemicals (Hazard Information and Packaging for Supply) Regulations 1994]

packaging Means, in relation to a substance or preparation dangerous for supply, as the context may require, the receptacle, or any components, materials or wrapping associated with the receptacle for the purpose of enabling it to perform its containment function or both.
[Chemicals (Hazard Information and Packaging for Supply) Regulations 1994]

part learning Learning, usually by way of memorizing, in which a task is divided into smaller units and each unit is separately learned.

passive fire fighting system (1C5) A fixed installation in a building comprising, for example, hydrants, dry risers and sprinkler systems supported by in situ smoke and heat detection systems.

paternoster lift A continuous running appliance for transporting passengers in a substantially vertical direction, in which a number of cars are suspended by two endless chains attached to the cars diagonally, so that the car floors are substantially horizontal when changing direction at the extremities of travel.

pathology The study of diseases.
The study of abnormal changes in the human body and their causes.

peak action level A level of peak sound pressure of 200 pascals.
(Noise at Work Regulations 1989)

penetrant (dye penetrant) inspection (2D4) A form of non-destructive testing where a crack in a metal is coated with oil containing a dye, followed by visual observation to ascertain the oil entering and leaving the crack under examination.

perception (1A2, 2A2) A general term referring to the awareness of objects, qualities or events stimulating a person's sense organs. Perception is the process of receiving information through the sensory channels – sight, hearing, touch, taste and smell.

perceptual defence (1A2) A feature of perception, this mechanism modifies, distorts or eliminates those stimuli that appear stressful, threatening or create anxiety.

perceptual distortion (1A2) This may take the form of perceptual defence and perceptual sensitization. See individual entries.

perceptual sensitization (1A2) People can become 'sensitized' to certain stimuli if they are, in their opinion, relevant or meaningful to them. Such stimuli are perceived as larger, brighter, more valuable and more attractive than other stimuli.

perceptual set (1A2) This is concerned with the processes of attention and interpretation which are closely related. Perceptual set consists of internal determinants, such as sensitivity, conscious or unconscious motives, attention span, culture and prejudice, and external determinants, including intensity, size, colour, location, senses and sense combinations, repetition, novelty and change. People have a perceptual set that selects, rejects, modifies, ignores or interprets that which is perceived.

permanent threshold shift A form of occupational deafness where the limit of tolerance is exceeded in terms of the duration of exposure to noise, the level of noise and individual susceptibility. Recovery from permanent threshold shift does not proceed to completion, effectively ceasing at some point in time after the end of the noise exposure. The condition is commonly known as 'permanent deafness'.

Permit to Work (1A3, 1C3) A form of safe system of work operated where there is a high degree of foreseeable risk.
A formal safety control system designed to prevent accidental injury to personnel, damage to plant, premises and particularly when work with a foreseeably high hazard content is undertaken and the precautions required are numerous and complex.

personal injury Includes any disease and any impairment of a person's physical or mental condition.
(Limitations Act 1980)

personality (1A2) The dynamic organization within the individual of those psychophysical systems that determine his or her characteristic behaviour and thought (Allport, 1961).
The distinctive characteristics of individuals, the stable and changing relationships between these characteristics, the origins of the characteristics, the ways in which they help or hinder the interaction of a person with other people, and the

characteristic ways in which a person thinks about himself (Morgan and King, 1971).

personality trait An aspect of personality that is reasonably characteristic of a person and distinguishes him in some way from other people.
A tendency to behave in a particular way according to one of the more stronger elements of an individual's personality.

personal protective equipment (PPE) PPE means all equipment (including clothing affording protection against the weather) which is intended to be worn or held by a person at work and which protects him or her against one or more risks to his health and safety, and any addition or accessory designed to meet this objective. (Personal Protective Equipment at Work Regulations 1992)

personal sampling (1E2) The process of measuring individual exposure to airborne contamination in the workplace by fitting a sampling device to those who may be exposed.
Personal sampling instruments include gas monitoring badges, impingers and filtration devices. See individual entries.

persuasive precedent A precedent which is not binding on a court but to which a judge may attach some importance.

pervasion A route of entry of hazardous substances and micro-organisms into the body and, in some cases, the blood stream, via the skin.

phagocytes See ***macrophages***.

phon The standard unit of loudness or the level of loudness.

photo-electric trip device This device features a curtain of light which can be arranged in either a horizontal or vertical configuration. Interruption of the curtain whilst the dangerous parts of machinery are moving results in a signal being given for the machine to stop. The speed of arrestment of moving parts should be such as to ensure that dangerous parts have come to rest before they can be reached by the operator.

photo-ionization detector A form of gas/vapour analysing device in which the contaminant is drawn into a cell and ionized by ultraviolet radiation. Contaminants dissociate into charged ions which generate a current between two electrodes mounted in a sample chamber. The current created is proportional to the contraction of contaminant present.

physical agents Those physical phenomena which may have an adverse effect on the body, e.g. heat, radiation, noise, vibration, pressure and humidity.

physical process operations (2C3) A range of process operations associated with:

- drying, blending and mixing of substances, crushing and grinding of solids;
- bulk transfer, of solids, liquids and gases.

physical risks The risks arising from exposure to physical phenomena such as excessive temperatures, noise and vibration, various particulates and radiation.

pitch A psychological attribute of tones, related to frequency, but not directly proportional to it.

Pitot static tube An instrument for the calculation of air velocity, particularly in ducts. The device comprises two concentric tubes with openings to measure total and static air pressures, from which the velocity can be calculated.

pitting A form of surface fatigue failure in metals arising from the formation of minute pits or indentations in the surface. It occurs as a result of the removal of small surface particles due to rubbing.

place of comparative or relative safety A fire resistive corridor/staircase in a very tall building to which occupants can go as a first resort before making their way to the open air safely when time allows.

place of safety Generally interpreted as a location in the open air where people can freely walk way from a building and not be affected by heat or smoke from a fire in that building.

planned motivation A technique by which the attitudes, and thereby the performance, of people can be improved; an industrial catalyst; a tool to maximize performance. Planned motivation schemes frequently form the basis for safety incentive schemes.

planned preventative maintenance programme (1D1, 2A3) Planned maintenance is maintenance organized and carried out with forethought, control and the use of records to a predetermined plan.

Planned preventative maintenance programmes should be documented to record:

- the item of plant, equipment, structure, etc. to be maintained, where possible by serial number;
- the maintenance procedure to be followed;
- the frequency of maintenance, e.g. daily, weekly, quarterly;
- the identity of the person responsible for ensuring the maintenance procedure has been completed satisfactorily;
- precautions necessary during maintenance activities.

plasticity (2D4) The ability of a material to stretch under stress. In tensile tests, materials behave in an elastic manner until the Yield Stress is reached, after which the material will not return to its original shape.

playback robot A form of robot which incorporates a memory. However, the robot has to be taught the sequence, positions and operations required by a human operator. When required, this information is recalled and the operations carried out automatically from memory.

Poisons Advisory Centre A body approved for the time being for the purposes of Regulation 14 (notification of constituents of certain preparations dangerous for supply) by the Secretary of State for Health in consultation with the Secretaries of State for Scotland and Wales, the Health and Safety Commission and such other persons or bodies as appear to him or her to be appropriate.

[Chemicals (Hazard Information and Packaging for Supply) Regulations 1994]

pollution of the environment Pollution of the environment due to the release (into any environmental medium), from any process, of substances that are capable of causing harm to human beings or any other living organisms supported by the environment.
(Environmental Protection Act 1990)

portable electrical equipment (appliances) (1D4) This includes all forms of portable electrical equipment, including portable electrical hand tools, lamps, kettles, etc. HSE Guidance Note PM 32 'The safe use of portable appliances' proves guidance on this matter, including recommendations for the regular testing and examination of such appliances. Electrical tests should confirm the integrity or otherwise of earthing and insulation.

portable fire fighting equipment (1C5) All forms of fire appliance which can be held in the hand and operated by hand.

positive pressure-powered dust respirator A form of dust respirator comprising an ori-nasal facepiece fitted to a power-driven pack carried on the individual and connected by a flexible hose.

positive reinforcement A form of reinforcement that the learner approaches readily.

Positive (Success) Tree Design An element of the Management Oversight and Risk Tree (MORT) system, this technique reverses the logic of fault tree analysis. In positive tree design, a system for successful operation is comprehensively and logically laid out. The positive tree is an excellent planning and assessment tool because it shows all that must be performed and the proper sequencing of events needed to accomplish an objective.

positive transfer of training More rapid learning in one situation because of previous learning in another situation which is similar in terms of the stimuli and/or responses required in the two situations.

postural fatigue A range of symptoms frequently experienced by users and operators of display screen equipment in particular, resulting in lumbar, arm and shoulder pain, due to the adoption of unsuitable postures at a workstation. The more abnormal the posture, the greater the potential for postural fatigue and even long-term injury.

potential difference An electric current will flow along a conductor if there is a difference in electrical pressure between the two ends. Potential difference is the difference in electrical pressure between these two points.

powered working platform A mobile working platform used where quick and safe access to plant, electrical installations and goods stored at high level is required. This form of platform operates from either a self-propelled electric trolley or a light trailer. Their operations take three specific modes: self-propelled hydraulic boom operation, semi-mechanized articulated boom operation and self-propelled scissor lift operation.

power interlocking A safety feature of machinery achieved by direct mechanical control of a switch in series with the mains power supply to the drive of the machinery. The direct mechanical control may be by links, captive key or trapped key.
This form of interlocking is commonly used for high risk situations.

power press A press or press brake for the working of metal by means of tools, or for die proving, which is power driven and which embodies a flywheel and clutch.
(Provision and Use of Work Equipment Regulations 1998)

practicable See ***hierarchy of duties***.

precedent A legal doctrine under which, in most cases, a court must follow the earlier decisions of courts at its own level and of the superior courts, known as *binding precedent*. Other precedents are of a *persuasive* nature and are not binding.

pre-demolition survey (2C4) A survey undertaken prior to demolition in order to ensure specific safety procedures during the demolition process.
A pre-demolition survey should identify:

- the nature and method of construction of the structure;
- the arrangement, and condition of, adjacent buildings and structures;
- the location of underground services;
- the previous use of the premises;
- the presence of dangerous substances inherent in the structure or stored;
- the method of bonding of load-bearing walls;
- the system for shoring or provision of other support during demolition;
- the presence of cantilevered structures;
- the presence of basements and other underground spaces;
- the presence of storage tanks above and below ground level;
- the actual sequence of operations for the demolition process, particularly if materials are to be recovered for future use.

pre-employment screening (1E3) A form of health surveillance aimed at assessing the general fitness of a person prior to commencing employment. As part of the screening process, it is standard practice in many organizations for intending employees to complete a Pre-Employment Health Screening Questionnaire for evaluation by an occupational health nurse or occupational physician.
Pre-employment health screening may incorporate a number of assessment techniques, such as vision screening, audiometry, fitness tests, measurement of weight and height, and a respiratory lung function test.

pregnant workers and nursing mothers (1C1) The Management of Health and Safety at Work Regulations 1999 require employers to:

- assess the risks to the health and safety of women who are pregnant, have recently given birth, or who are breast feeding; and

- ensure that workers are not exposed to risks identified by the risk assessment, which would present a danger to their health or safety.

If, after taking whatever preventive action is reasonable, there is still a significant risk, which goes beyond the level of risk to be expected outside the workplace, the employer must take the following steps to remove the employee from that risk:

- temporarily adjust working conditions or hours of work;
- if that is not reasonable or would not avoid the risk, offer alternative work if any is available;
- if that is not possible, give her paid leave from work for as long as is necessary to protect her health and safety or that of her child.

If a new or expectant mother works at night and has a medical certificate stating that night work could damage her health or safety, the employer must either offer her daytime work if any is available or, if that is not reasonable, give her paid leave for as long as necessary to protect her health or safety

preparation A mixture or solution of two or more substances.
(Control of Substances Hazardous to Health Regulations 1999)

preparation dangerous for supply A preparation is dangerous for supply if it falls within at least one of the categories of danger specified in Column 1 of Schedule 1 to the Chemicals (Hazard Information and Packaging for Supply) Regulations 1994.

presbycusis Deafness resulting from the ageing process, in particular the reduction in hearing acuity in the case of frequencies at the higher end of the frequency range.

prescribed disease A disease may be prescribed if:

(a) it ought to be treated, having regard to its causes and incidence and other relevant considerations, as a risk of occupation and not a risk common to all persons; and
(b) it is such that, in the absence of special circumstances, the attribution of particular cases to the nature of the employment can be established with reasonable certainty.

[Social Security Act 1975, section 76(2)]
Current requirements relating to prescribed occupational diseases are covered by the Social Security (Industrial Injuries)(Prescribed Diseases) Regulations 1985 and various amendments to these regulations.

prescribed process A process prescribed under section 2(1) of the Act.
(Environmental Protection Act 1990)

pressure gauge A safety device which indicates the pressure of steam in a boiler. The maximum working pressure should be clearly marked on the gauge glass.

pressure-sensitive mat A safety device operating by means of a number of suitably placed electrical or fluid switches/valves contained within a mat connected to a control unit and covering the approaches to a danger area of a machine or other installation. Pressure on the mat operates one or more of the switches, causing the machine to stop.

pressure systems (2D5) This means:

(a) a system comprising one or more pressure vessels of rigid construction, any associated pipework and protective devices;
(b) the pipework with its protective devices to which a transportable gas container is, or is intended to be, connected; or
(c) a pipeline and its protective devices.

(Pressure Systems Regulations 1999)

pre-trial review A review of a claim in the County Court undertaken by the Registrar with the objective of identifying the key issues of the claim and any points of contention.

'preventive and protective measures' The measures identified by an employer or self-employed person, in consequence of the (risk) assessment, as the measures he or she needs to take to comply with the requirements and prohibitions imposed upon him or her by or under the relevant statutory provisions and by Part II of the Fire Precautions (Workplace) Regulations 1999.
(Management of Health and Safety at Work Regulations 1999)

primary irritant A substance which can cause dermatitis principally at the site of contact if exposed for a sufficient length of time and in sufficient concentrations, e.g. acids, alkalis and solvents.

primary monitoring An area of occupational health practice dealing with the clinical observation of sick people who may seek advice and/or treatment.

prime mover Every engine, motor or other appliance which provides mechanical energy derived from steam, water, wind, electricity, the combustion of fuel or other source.
(Factories Act 1961)

Principle of Definition (Total Loss Control) A logical decision can be made only if the real problem is first defined.

Principle of Future Characteristics (Total Loss Control) An organization's past performance tends to foreshadow its future characteristics.

Principle of Management Results (Total Loss Control) A loss control programmer tends to secure most effective results through and with others by planning, organizing, leading and controlling.

Principle of Moments When a body is in equilibrium under the action of any number of forces acting in a plane, the sum of the moments of those forces about any point in the plane is zero.

Principle of Multiple Causes (Total Loss Control) A loss is seldom, if ever, the result of a single cause.

Principle of Point of Control (Total Loss Control) The greatest potential for control tends to exist at the point where the action takes place.

Principle of Reciprocated Interest (Total Loss Control) People tend to be motivated to achieve the results that you want to the extent that you show interest in the results *they* want to achieve.

Principle of Recognition (Total Loss Control) Motivation to accomplish results tends to increase as people are given recognition for their contribution to those results.

Principle of Reporting to the Highest Authority (Total Loss Control) The higher the level to which a loss control manager reports, the more management co-operation he achieves.

Principle of Resistance to Change (Total Loss Control) The greater the departure of planned changes from the accepted ways, the greater the potential resistance by the people involved.

Principle of the Critical Few (Total Loss Control) In any given group of occurrences, a small number of causes will tend to give rise to the largest proportion of results.

private nuisance (1B1) Some form of act, failure to act or omission on the part of an individual or group that results in inconvenience, damage or obstruction to another individual or group.

proactive measures (1A5) Proactive measures for determining standards of health and safety performance include a number of health and safety monitoring techniques, including occupational health surveys by an occupational health nurse, safety audits, safety inspections, safety sampling exercises and safety surveys.
It is important that any health and safety objectives agreed with managers to improve health and safety performance are both measurable and achievable by the persons concerned, in terms of the time and resources available.

probabilistic problems In machinery maintenance operations, the timing and outcome of maintenance may depend upon chance. From a frequency distribution of the time elapsed between maintenance operations and failure, it is possible to determine the variations in the probability of failure with elapsed time.

process Any activities carried out in Great Britain, whether on premises or by means of mobile plant, that are capable of causing pollution of the environment. (Environmental Protection Act 1990)

producer In relation to a product means the following:

- the person who manufactured it;

- in the case of a substance that has not been manufactured, but has been won or abstracted, the person who won or abstracted it;
- in the case of a product that has not been manufactured, won or abstracted, for which the essential characteristics are attributable to an industrial process having been carried out, such as in relation to agricultural produce, the person who has carried out that process.

(Consumer Protection Act 1987)

product Any goods or electricity and (subject to the proviso below) includes products that are used in another product, whether by virtue of being a component part or raw material for it or otherwise. The proviso is that a person who supplies any product in which products are used, whether by virtue of being component parts or raw materials or otherwise, shall not be treated by reason only of their supply of that product, but as supplying any of the products so comprised.
(Consumer Protection Act 1987)

Product Liability (2B4) An area of health and safety law concerned with both the criminal and civil liabilities of all those in the manufacturing chain towards consumers of their products. Criminal liability is covered in the HSWA (section 6) and other ancillary legislation, such as the Consumer Protection Act 1987. Injury sustained as a result of using a defective product could result in a civil claim against a defendant based on negligence.
Principal duties rest with designers, manufacturers and importers of products, secondary duties with wholesalers, retailers and other persons directly or indirectly involved in the supply chain.

programmed instruction/learning A form of instruction/teaching in which the following factors are present:

- there is a clear statement of exactly what the trainee is expected to be able to do at the end of the programme;
- the material to be learned, which is itemized and tested, is presented in identifiable steps and/or frames;
- trainees follow an actual sequence of frames that is determined for them according to their individual needs;
- feedback of the information of the correctness or otherwise of responses is usually given to the trainee before the next frame is presented.

programmed learning A form of self-instruction by means of carefully designed questions or items which, through immediate reinforcement, motivate and enhance the learning process.

progressive chronic effect An adverse effect on health which continues to develop after exposure to an offending agent has ceased.

prohibition A control strategy in accident and ill health prevention exercised where there is no known form of operator protection available. This may entail prohibiting the use of a substance, system of work, operational practice or machine where the level of danger is very high.

Prohibition Notice Enforcement officers are empowered to serve Prohibition Notices on certain circumstances thus:

- Where there is considered to be a serious risk of injury to employees and visitors from fire on the premises, the fire authority can serve a Prohibition Notice on the occupier requiring that remedial work be carried out in the interests of fire safety or, alternatively, have the premises closed down.
  (Fire Safety and Safety of Places of Sport Act 1987)
- Where the enforcing authority is of the opinion, as respects the carrying out of a prescribed process under an authorization, that to continue to carry it out or to do so in a particular manner involves an imminent risk of serious pollution of the environment, the authority shall serve a Prohibition Notice on the person carrying out the process.
  (Environmental Protection Act 1990)
- Where an inspector is of the opinion that a work activity involves or will involve a risk of serious personal injury, he or she may serve a Prohibition Notice on the owner and/or occupier of the premises or the person having control of that activity.
  (Health and Safety at Work etc. Act 1974)

prohibition sign A circular sign with a red band enclosing a crossed-out symbol on a white background, e.g. 'No Smoking'.
(Safety Signs Regulations 1980)

project A training activity directed at requiring the trainee to put into practice information acquired during training sessions. A project incorporates objectives which must be met, together with guidelines for meeting these objectives.
A project which includes or is intended to include construction work.
[Construction (Design and Management) Regulations 1994]

Project Safety Analysis A joint exercise undertaken by a group of specialists (e.g. engineer, architect, liability surveyor and health and safety specialist), with the aim of ensuring that account is taken of accumulated experience, knowledge of the technology and best practice in the initial design of a project.

prolapsed intervertebral disc A prolapsed or 'slipped' disc occurs when one of the intervertebral discs is displaced from its normal position and is no longer performing its function properly. In other cases, there may be squashing or compression of the disc.
Slipped discs are commonly associated with incorrect manual handling techniques.

proof load A force applied in a static tensile test of lifting tackle (e.g. chains, slings), to which every item must be subjected without indicating any defects on visual examination. Proof loads are commonly set at twice the safe working load.

proof stress The amount of stress produced by a certain amount of permanent extension in a material under test conditions.

prosecution The bringing of a person before a court to answer a charge involving an alleged breach of the law. It is normal for the person charged with an offence to be served with a summons to attend court to answer the charge or charges in question.

protected route An escape route with fire-resisting construction and free of fire risks that will protect people from smoke and heat whilst they make their way to a place clear of risk.

public nuisance (1B1) This form of nuisance has a direct effect on the public at large, such as the inconvenience arising from the obstruction of a public right of way.

Pure Chance Theory This theory of accident causation suggests that everyone has an equal chance of having an accident. Accidents are viewed as an act of God, over which the accident victim has no control, and suggesting that accident prevention measures are futile.

putlog (tied-in) scaffold A scaffold incorporating a single row of standards, connected to an existing building, or erected as a building progresses upwards, by installing the flattened end or putlog of the transom (cross member) into a vertical joint between two bricks.

Q

quality (2A3) The totality of features and characteristics of a product or service that bear on its ability to satisfy stated or implied needs.
(BS 4778: *Quality Vocabulary*)

Quality Assurance (2A3) All those planned and systematic actions necessary to provide adequate confidence that a product or service will satisfy given requirements of quality.
(BS 4778: *Quality Vocabulary*)

quality audit A form of audit undertaken to verify whether the quality system is being followed as written and whether the system itself meets the requirements of the customer.
(BS 4778: *Quality Vocabulary*)

quality control (2A3) A series of operational techniques and activities used to fulfil and verify the requirements for quality; for example, inspection and examination systems.
(BS 4778: *Quality Vocabulary*)

Quality of Working Life This philosophy takes into account the various theories of motivation and worker satisfaction, placing them in a wider framework. It takes into account a range of factors that affect the basic satisfaction of workers, such as a well-organized working environment and the maintenance of adequate welfare facilities, and hence, the quality of their working life.
This approach on the 1974 International Labour Office's conference resolution directed at promoting:

- protection against physical conditions and dangers at the workplace and its immediate environment;
- adaptation of installations and work process to the physical and mental aptitudes of the worker through the application of ergonomic principles;
- prevention of mental stress due to the pace and monotony of work and the promotion of the quality of working life through amelioration of the conditions of work, including job design and job content and related questions of work organization;
- the full participation of employers and workers and their organizations in the elaboration, planning and implementation of policies for the improvement of the working environment.

quarry A system of excavations for minerals which is not a mine.

(Mines and Quarries Act 1954)

quartz fibre detector (packet electrometer) A device for detecting radiation which comprises a metal cylinder with a loop of quartz fibre in the middle which is charged to approximately 200 volts. Electrical attraction or repulsion takes place between the fibre and the outside case. Ionizing radiation allows the charge to leak away slowly, permitting the loop to move over a scale back to the centre of the cylinder. The device is very accurate, measuring down to 5 millirem of dose.

R

radiation (1E1) The process of heat transfer through the air or gases.
Radiation is a form of energy and sources of radiation can be classified as non-ionizing and ionizing radiation.
Ionizing radiation can produce chemical changes in the body as a result of ionizing molecules upon which it is incident. Non-ionizing radiation does not have this effect and is usually absorbed by the molecules on which it is incident, with the result that the material will heat up, as in the case of microwaves.

radiation dose (2E3) The quantity of radiation in relation to the duration of exposure.

radiation dose equivalent The quantity obtained by multiplying the absorbed dose of radiation by a factor to allow for the differing effectiveness of the various ionizing radiations in causing harm to tissue. The unit is the sievert.

radioactivity (2E3) The process of spontaneous disintegration of heavy elements, with the emission of particles and energy, and changing into other elements in the process.

radiography (2D4) A non-destructive testing technique aimed at detecting internal flaws in materials and using gamma rays, neutrons and X-rays. Radiographic techniques are concerned with detecting the thickness of metals and the degree of fusion of welds on pipes and internal surfaces.

raker In tubular steel scaffolding, an inclined load-bearing tube attached to the main body of a scaffold. It may bear on the ground or an adjacent structure.

ratio decidendi One of the features of a judgement, meaning 'the reason for deciding', and is the legal principle or rule behind a particular judgement, the actual finding on a particular fact. It is essentially a proposition of law that decides the case in the light of, or in the context of, the material facts.

reactive measures/monitoring (1A5) These are post-incident measures directed at preventing a recurrence of accidents and occupational ill health. They include a range of techniques, such as:

- accident investigation;
- the production and interpretation of statistical information;
- the provision of information, instruction, training and supervision;
- health surveillance;
- air monitoring;

- the provision and use of personal protective equipment;
- specific protection for vulnerable groups, e.g. young persons, pregnant workers.

'reasonable' People are expected to behave in a reasonable way towards each other. Hence the concept of the mythical 'reasonable man', interpreted by one judge in the past as comparable with 'the man who travels to work every day on the top deck of the No. 57 Clapham omnibus'. The term is flexible and changes with the norms of society at the time.
Under the HSWA, employees are required to take 'reasonable care' for the safety of themselves and others who may be affected by their acts or omissions.

reasonable care (2E6) A rule of common law that everyone owes a duty to everyone else to take 'reasonable care' so as not to cause them foreseeable injury. Employers must take reasonable care to protect their employees from risk of foreseeable death, injury or disease at work.

reasonably practicable A qualified duty implying a lesser level of duty than a duty qualified by the term 'so far as is practicable'.
'Reasonably practicable' is a narrower term than 'physically possible' and implies that a computation must be made in which the quantum of risk is placed on one side of the scale and the sacrifice involved in carrying out the measures necessary for averting the risk is placed on the other side. If it can be shown there is gross disproportion between these two factors, i.e. the risk is insignificant in relation to the sacrifice, then a defendant discharges the onus on themselves.
[Edwards v. National Coal Board (1949) 1 AER 743]
See ***hierarchy of duties***.

receptacle This means, in relation to a substance or preparation dangerous for supply, a vessel, or the innermost layer of packaging, which is in contact with a substance and which is liable to be individually handled when the substance is used, and includes any closure or fastener.
[Chemicals (Hazard Information and Packaging for Supply) Regulations 1994]

receptor system A form of local exhaust ventilation system whereby the contaminant enters the system without inducement, being transported from a hood by means of a fan and ducting to a collection unit.
There are three types – a total enclosure, such as a laboratory fume cupboard; partial enclosure, such as a spray booth; and a receptor hood, which may be directly above a source of contaminant.
Hoods which receive the contaminated air as it flows from its origin under the influence of thermal currents are generally classed as receptors. See Figure R.1.

reciprocating trap A form of machinery hazard associated with up and down or sideways movement. At the point where the injury occurs, the limb is stationary. See Figure R.2.

recognized trade union An independent trade union, as defined in section 30(1) of the Trade Union and Labour Relations Act 1974, that the employer concerned recognizes for the purpose of negotiations relating to or connected with one or

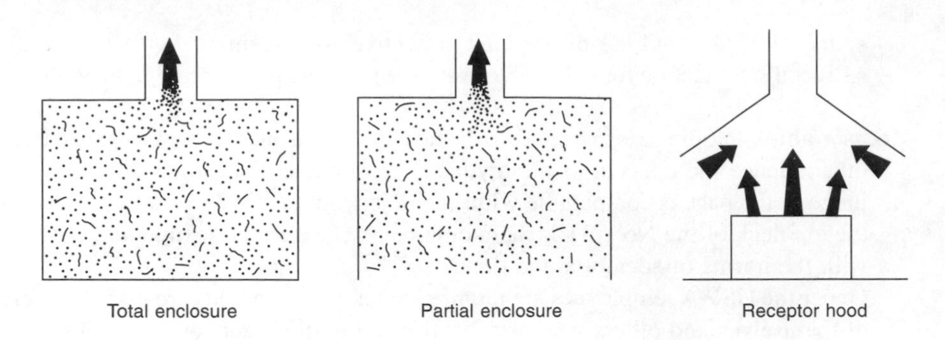

Figure R.1 Receptor systems.

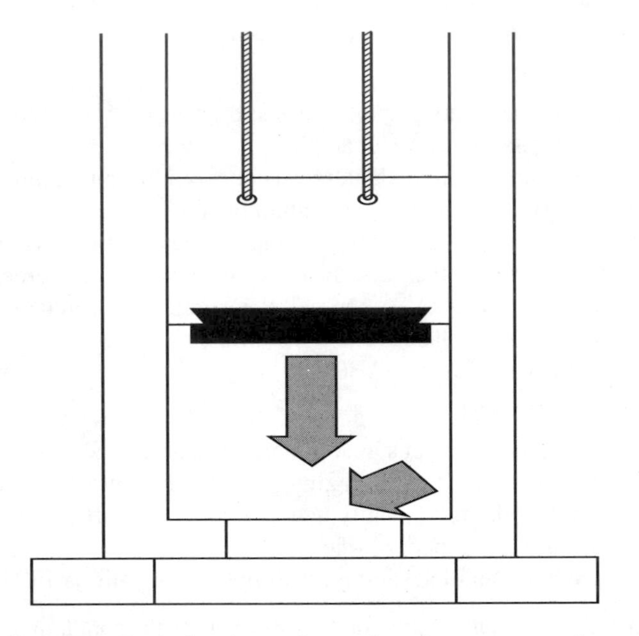

Figure R.2 Vertical reciprocating motion of a power press. Source: BS 5304 (1988) Safeguarding of Machinery.

more of the matters specified in section 29(1) of the Act regarding those employed by them or one that the Advisory, Conciliation and Arbitration Service has recommended to be recognized under the Employment Act 1975, which is operative within the meaning of section 15 of the Act.
(Safety Representatives and Safety Committees Regulations 1977)

reduced time exposure (limitation) (2E2) A strategy directed to limiting the exposure of people to, for instance, noise or hazardous substances, by specifying the maximum exposure time permissible in any working period, e.g. 8 hours. This strategy forms the basis for long-term and short-term exposure limits.

reduced voltage A form of protection against electric shock, the most commonly used reduced voltage system being the 110-volt centre point earthed system. Here, the secondary winding of the transformer providing the 110-volt supply is centre tapped to earth, thereby ensuring that at no part of the 110-volt system can the voltage to earth exceed 55 volts.

reflected glare The reflection of bright light sources on shiny or wet work surfaces, such as glass or plated metal, which can almost entirely conceal the detail in or behind the object which is glinting.

regulated stand A stand providing covered accommodation for 500 or more spectators.
(Fire Safety and Safety of Places of Sport Act 1987)

Regulations (1B1, 1B2, 2B2) The HSWA is an enabling Act. Section 15, Schedule 3 and section 80 of the HSWA empower the Secretary of State for Employment to make Statutory Instruments which specify more detailed rules and requirements for implementing the broad aims of the Act with regard to, for instance, noise and electrical safety. These take the form of Regulations and are termed 'delegated' or 'subordinate' legislation.
Regulations may:

- repeal or modify existing statutory provisions;
- exclude or modify the provisions of sections 9 of the HSWA or existing provisions in relation to any specified class;
- make a specified authority, such as the HSE or a local authority, responsible for enforcement;
- impose approval requirements;
- refer to specified documents to operate as references;
- give exemptions;
- specify the class of person who may be guilty of an offence, e.g. employers;
- apply to a particular case only.

Examples of Regulations are:

- Management of Health and Safety at Work Regulations 1999.
- Provision and Use of Work Equipment Regulations 1998.
- Control of Substances Hazardous to Health Regulations 1999.

Most Regulations now implement the requirements of European Directives.

reinforcement (2A2) A stimulus or event that strengthens a response when it follows the response.

relative humidity (1E1) This is the amount of moisture present in the air at a particular time compared with that which would produce total saturation. Relative humidity is expressed as a percentage.
Relative humidity is an important feature of comfort in the working environment. High levels of humidity produce physical discomfort and excessive perspiration. On the other hand, low levels cause dryness of the throat and nasal passages. To

ensure reasonable levels of comfort in a workplace, relative humidity should be around 70%.

released A substance is released into an environmental medium whenever it is released directly into that medium, whether or not it is released into it within or outside Great Britain, and release includes:

- in relation to air, any emission of the substance into the air;
- in relation to water, any entry (including any discharge) of the substance into water;
- in relation to land, any deposit, keeping or disposal of the substance in or on land.

(Environmental Protection Act 1990)

Relevant Enforcing Authority In the case of lifting equipment:

(a) where the defective lifting equipment has been hired or leased by the employer, the HSE;
(b) otherwise, the enforcing authority for the premises in which the defective lifting equipment was thoroughly examined.

(Lifting Operations and Lifting Equipment Regulations 1998)

relevant statutory provisions (1B1) The HSWA is an umbrella or enabling Act comprising general duties only. Section 15 of the HSWA enables the Secretary of State for Employment to make Regulations. The Act and any regulations made under the Act are deemed to be the relevant statutory provisions.
Schedule 1 of the HSWA lists regulations deemed to be the relevant statutory provisions.
The relevant statutory provisions include:

- Part 1 of the HSWA;
- regulations made under Part 1;
- the Acts contained in Schedule 1 of the HSWA, e.g. Factories Act 1961;
- any regulations made under the above Acts.

relevant transport system A railway, tramway, trolley vehicle system or guided transport system.
(Reporting of Injuries, Diseases and Dangerous Occurrences Regulations 1995)

reliability The ability of an item to perform a required function under stated conditions for a stated period of time.
(BS 4774)
The self-consistency of a method of measurement, or the degree to which separate independent measurements of the same thing agree with each other. Reliability is expressed by a coefficient of correlation representing a relationship between the two sets of measurements of the same thing.

reliability data Data produced as a result of the measurement of reliability, generally to support a particular statement supporting that which was measured.

reliability theory A management tool based on the quantitative assessment of reliability designed to maximize efficiency by optimizing performance.

repetitive strain injury (RSI) This condition covers well-known conditions such as tennis elbow, flexor tenosynovitis and carpal tunnel syndrome. It is commonly caused or aggravated by work associated with repetitive and over-forceful movement, excessive workloads, inadequate rest periods and sustained or constrained postures. This results in pain or soreness due to inflammation of the muscles and the synovial lining of the tendon sheath.
Signs and symptoms of RSI include localized pain, tenderness, swelling and a grating sensation in the joint (crepitus) aggravated by pressure or movement.

replacement problems In maintenance programmes, these problems are concerned with determining the best time at which an item, such as a component or a more complex machine, should be replaced. This may entail a complete overhaul of machinery or the more simple replacement of worn parts.

report A written record of activities based on authoritative sources, written by a qualified person and directed towards a predetermined group.

reportable disease (1A4) Under the Reporting of Injuries, Diseases and Dangerous Occurrences Regulations (RIDDOR) 1995, certain diseases and conditions arising from work activities affecting a person at work, and listed in Schedule 3, must be reported by an employer to the relevant enforcing authority.
Reportable diseases are classified in the following groups:

1. Conditions due to physical agents and the physical demands of work.
2. Infections due to biological agents.
3. Conditions due to substances.

See Figure R.3.

reportable event (1A4) The Reporting of Injuries, Diseases and Dangerous Occurrences Regulations (RIDDOR) 1995 apply to 'events' which arise out of or in connection with work, namely all deaths, certain injuries resulting from accidents, instances of specified diseases and defined dangerous occurrences.
Where any of the events listed below arise out of work activities, it must be notified by quickest practicable means (e.g. telephone or fax), and subsequently reported (within 10 days) on the appropriate form, to the enforcing authority.
The events are:

- the death of any person at work as a result of an accident, whether or not they are at work;
- someone who is at work suffering a major injury as a result of an accident arising out of or in connection with work;
- someone who is not at work (e.g. a member of the public) suffers an injury as

Health and Safety at Work etc Act 1974
The Reporting of Injuries, Diseases and Dangerous Occurrences Regulations 1995

Report of a case of disease

Filling in this form
This form must be filled in by an employer or other responsible person.

Part A

About you

1 What is your full name?

2 What is your job title?

3 What is your telephone number?

About your organisation

4 What is the name of your organisation?

5 What is its address and postcode?

6 Does the affected person usually work at this address?

Yes ☐ Go to question 7

No ☐ Where do they normally work?

7 What type of work does the organisation do?

Part B

About the affected person

1 What is their full name?

2 What is their date of birth?

/ /

3 What is their job title?

4 Are they

☐ Male?

☐ Female?

5 Is the affected person (tick one box)

☐ One of your employees?

☐ On a training scheme? Give details:

☐ On work experience?

☐ Employed by someone else? Give details:

☐ Other? Give details:

Figure R.3 Report of a case of disease.

Part C

The disease you are reporting

1 Please give:
- The name of the disease, and the type of work it is associated with; or
- The name and number of the disease (from Schedule 3 of the Regulations– see the accompanying notes).

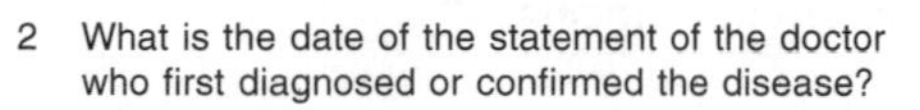

2 What is the date of the statement of the doctor who first diagnosed or confirmed the disease?

/ /

3 What is the name and address of the doctor?

Part D

Describing the work that led to the disease

Please describe any work done by the affected person which might have led to them getting the disease.

If the disease is thought to have been caused by exposure to an agent at work (eg a specific chemical) please say what that agent is.

Give any other information which is relevant.

Give your description here

Continue your description here

Part E

Your signature

Signature

Date

/ /

Where to send the form

Please send it to the Enforcing Authority for the place where the affected person works. If you do not know the Enforcing Authority, send it to the nearest HSE office.

For official use

Client number

Location number

Event number

☐ Inv Rep ☐ Y ☐ N

Figure R.3 Cont'd

a result of an accident and is taken from the scene to a hospital for treatment, or if the accident happens at a hospital, suffers a major injury;

- one of the list of specified dangerous occurrences takes place;
- someone at work is unable to do their normal work for more than 3 days as a result of an injury caused by an accident at work;
- the death of an employee, if this occurs some time after a reportable injury which led to the employee's death, but not more than 1 year afterwards;
- a person at work suffers one of a number of specified diseases provided that a doctor diagnoses the disease and the person's job involves a specified work activity.

The duty to notify and report rests with the responsible person, i.e. the employer, a self-employed person or person in control of the premises.
A report must be made of the following circumstances:

- a conveyor of flammable gas through a fixed pipe distribution system, or the filler, importer or supplier of liquefied petroleum gas (LPG) in a refillable container, must report if they learn that someone has died or suffered a major injury arising out of, or in connection with, that gas;
- any registered installation business must report if it finds that there is, in any premises, a gas fitting or associated flue or ventilation arrangement which could be dangerous.

A responsible person must keep a record of the above events.
There is a defence available for a person to prove that he or she was not aware of the event requiring him or her to notify or send a report to the relevant authority, and that he or she had taken all reasonable steps to have such events brought to his notice.

reportable major injury (1A4) Under the Reporting of Injuries, Diseases and Dangerous Occurrences Regulations (RIDDOR) 1995, reportable major injuries are classified thus:

- Any fracture, other than to the fingers, thumbs or toes.
- Any amputation.
- Dislocation of the shoulder, hip, knee or spine.
- Loss of sight (whether temporary or permanent).
- A chemical or hot metal burn to the eye or any penetrating injury to the eye.
- Any injury resulting from electric shock (including any electrical burn caused by arcing or arcing products) leading to unconsciousness or requiring resuscitation or admittance to hospital for more than 24 hours.
- Any other injury:
  - leading to hypothermia, heat induced illness or to unconsciousness;
  - requiring resuscitation; or
  - requiring admittance to hospital for more than 24 hours.
- Loss of consciousness caused by asphyxia or by exposure to a harmful substance or biological agent.

- Either of the following conditions which result from the absorption of any substance by inhalation, ingestion or through the skin:
  - acute illness requiring medical treatment; or
  - loss of consciousness.
- Acute illness which requires medical treatment where there is reason to believe that this resulted from exposure to a biological agent or its toxins or infected material.

See Figure R.4.

residual current device See ***earth leakage circuit breaker***.

res ipsa loquitur A term meaning 'the facts speak for themselves', commonly used in civil actions. The term implies that the negligence in the particular case is self-evident. Had the claimant taken proper precautions, the injury, damage or loss would not have resulted.

resistance Ohm's law states that, at a constant temperature, the current through a conductor is directly proportional to the potential difference across its ends.

The ohm is the practical unit of resistance, being that resistance which allows a current of 1 amp to flow through it when the potential difference across it is 1 volt.

$$\text{Resistance} = \frac{\text{Pressure}}{\text{Current}} = \frac{\text{Volts}}{\text{Amps}}$$

respirable dust Airborne material which is capable of penetrating to the gas exchange region of the lung.
(Control of Substances Hazardous to Health Regulations 1999)

respiratory irritant A toxic substance which can have an immediate or delayed effect on the respiratory system. Acute exposure may cause pulmonary oedema.

respiratory protection (1E4) Respiratory protection entails the use of certain types of respirator designed to protect the respiratory system of the wearer against harmful dusts, gases, fumes and other forms of airborne contamination. The correct selection of respiratory protective equipment is essential. Employers must decide on the respiratory protection required in individual circumstances on the basis of a formal risk assessment required under the Personal Protective Equipment at Work Regulations 1992.
Respiratory protection may take the form of:

- general purpose dust respirators;
- positive pressure powered dust respirators;
- helmet-contained positive pressure respirators;
- gas respirators: cartridge type or canister type;
- emergency escape respirators.

See individual entries.

Health and Safety at Work etc Act 1974
The Reporting of Injuries, Diseases and Dangerous Occurrences Regulations 1995

Report of an injury or dangerous occurrence

Filling in this form
This form must be filled in by an employer or other responsible person.

Part A

About you

1 What is your full name?

2 What is your job title?

3 What is your telephone number?

About your organisation

4 What is the name of your organisation?

5 What is its address and postcode?

6 What type of work does the organisation do?

Part B

About the incident

1 On what date did the incident happen?

/ /

2 At what time did the incident happen?
(Please use the 24-hour clock eg 0600)

3 Did the incident happen at the above address?

Yes ☐ Go to question 4

No ☐ Where did the incident happen?

☐ Elsewhere in your organisation – give the name, address and postcode

☐ At someone else's premises – give the name, address and postcode

☐ In a public place – give details of where it happened

If you do not know the postcode, what is the name of the local authority?

4 In which department, or where on the premises, did the incident happen?

Part C

About the injured person
If you are reporting a dangerous occurrence, go to Part F.
If more than one person was injured in the same incident, please attach the details asked for in Part C and Part D for each injured person.

1 What is their full name?

2 What is their home address and postcode?

3 What is their home phone number?

4 How old are they?

5 Are they

☐ Male?

☐ Female?

6 What is their job title?

7 Was the injured person (tick only one box)

☐ One of your employees?

☐ On a training scheme? Give details:

☐ On work experience?

☐ Employed by someone else? Give details of the employer:

☐ Self-employed and at work?

☐ A member of the public?

Part D

About the injury

1 What was the injury? (eg fracture, laceration)

2 What part of the body was injured?

Figure R.4 Report of an injury or dangerous occurrence.

3 Was the injury (tick the one box that applies)

- ☐ A fatality?
- ☐ A major injury or condition? (see accompanying notes)
- ☐ An injury to an employee or self-employed person which prevented them doing their normal work for more than 3 days?
- ☐ An injury to a member of the public which meant they had to be taken from the scene of the accident to a hospital for treatment?

4 Did the injured person (tick all the boxes that apply)

- ☐ Become unconscious?
- ☐ Need resuscitation?
- ☐ Remain in hospital for more than 24 hours?
- ☐ None of the above.

Part E

About the kind of accident

Please tick the one box that best describes what happened, then go to Part G.

- ☐ Contact with moving machinery or material being machined
- ☐ Hit by a moving, flying or falling object
- ☐ Hit by a moving vehicle
- ☐ Hit something fixed or stationary

- ☐ Injured while handling, lifting or carrying
- ☐ Slipped, tripped or fell on the same level
- ☐ Fell from a height
  How high was the fall?
  ☐ metres

- ☐ Trapped by something collapsing

- ☐ Drowned or asphyxiated
- ☐ Exposed to, or in contact with, a harmful substance
- ☐ Exposed to fire
- ☐ Exposed to an explosion

- ☐ Contact with electricity or an electrical discharge
- ☐ Injured by an animal
- ☐ Physically assaulted by a person

- ☐ Another kind of accident (describe it in Part G)

Part F

Dangerous occurrences

Enter the number of the dangerous occurrence you are reporting. (The numbers are given in the Regulations and in the notes which accompany this form.)

☐

Part G

Describing what happened

Give as much detail as you can. For instance

- The name of any substance involved
- The name and type of any machine involved
- The events that led to the incident
- The part played by any people.

If it was a personal injury, give details of what the person was doing. Describe any action that has since been taken to prevent a similar incident. Use a separate piece of paper if you need to.

Part H

Your signature

Signature

Date

/ /

Where to send the form

Please send it to the Enforcing Authority for the place where it happened. If you do not know the Enforcing Authority, send it to the nearest HSE office.

For official use

Client number	Location number	Event number	
			☐ Inv Rep ☐ Y ☐ N

Figure R.4 Cont'd.

respiratory system (1E1) A vital body system whereby air, which contains oxygen, is brought into the body during the process of inhalation and its by-products, such as carbon dioxide, expelled during the process of exhalation. Oxygen is taken up by the haemoglobin in the blood and the red blood cells discharge carbon dioxide back into the alveoli to be exhaled.

The respiratory system includes the lungs and diaphragm and the upper respiratory tract, i.e. the nose and mouth, larynx, pharynx and trachea. See Figure R.5.

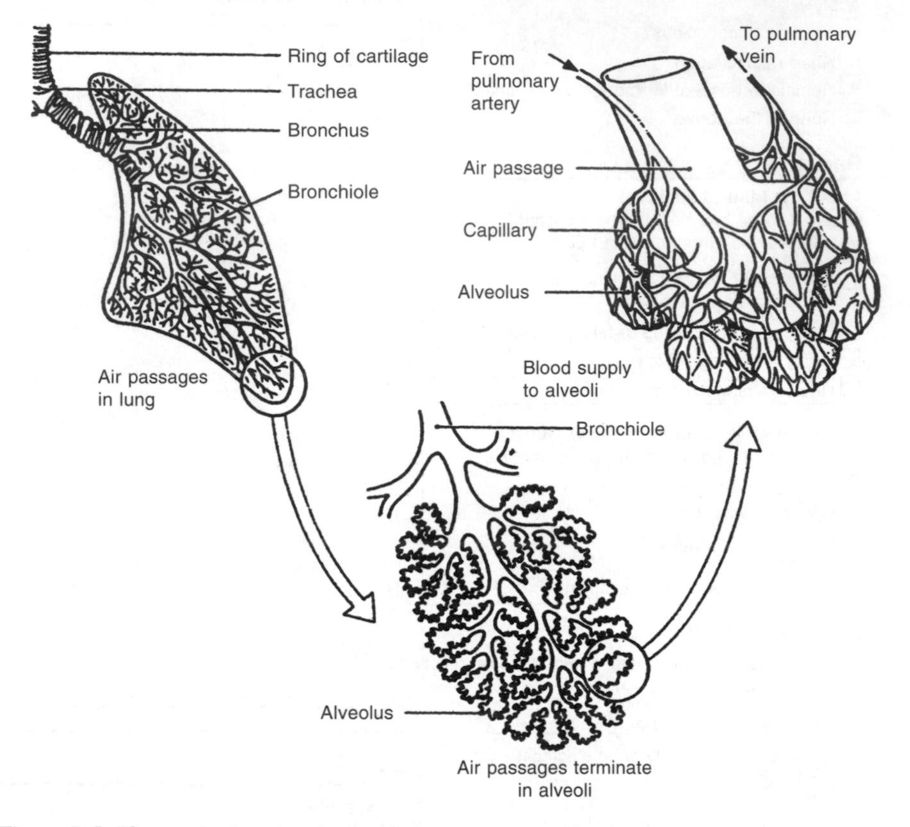

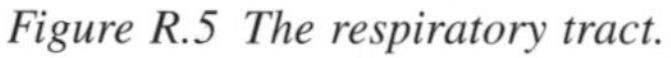

Figure R.5 The respiratory tract.

respirator zone A designated zone in any zone where the concentration of asbestos exceeds, or is liable to exceed, the control limit.

Respirator zones must be specifically demarcated and identified by notices and the notices must require employees entering the zone to wear respiratory protective equipment.

(Control of Asbestos at Work Regulations 1987)

retention (2A3) The amount correctly remembered; retention is measured on the basis of recognition, recall and savings.

retrospective studies A form of epidemiological study used to determine whether

there is a causal link between exposure to a substance and a certain level of diagnosed cases of diseases in a particular area, e.g. bladder cancer.
Such studies use two groups or cohorts, namely those who have been exposed and an unexposed control group.

risk (1A1) Risk expresses the likelihood or probability that the harm from a particular hazard will be realized.

risk assessment (1A1) The Management of Health and Safety at Work Regulations 1999 place an absolute duty on every employer to undertake a suitable and sufficient assessment of:

- the risks to the health and safety of employees to which they are exposed whilst at work; and
- the risks to the health and safety of persons not in their employment arising out of or in connection with the conduct by them of their undertaking.

This is for the purpose of identifying the measures he or she needs to take to comply with the requirements and prohibitions imposed upon him or her by or under the relevant statutory provisions.
A suitable and sufficient risk assessment should:

- identify the significant risks arising out of the work;
- enable the employer to identify and prioritize the measures that need to be taken to comply with the relevant statutory provisions;
- be appropriate to the nature of the work and such that it remains in force for a reasonable period of time.

In particular, a risk assessment should:

- ensure all relevant risks are addressed;
- address what actually happens in the workplace or during the work activity;
- ensure that all groups of employees and others who might be affected are considered;
- identify groups of workers who may be particularly at risk;
- take account of existing preventive or precautionary measures.

The significant findings should include:

- the significant hazards identified in the assessment;
- the existing control measures in place and the extent to which they control the risk;
- the population which might be affected by these significant risks or hazards, including any groups of employees who are especially at risk.

The ACOP to the above regulations states:

A risk assessment should usually involve identifying the hazards present in any undertaking (whether arising from work activities or from other factors) and then evaluating the extent of the risks involved, taking into account whatever precautions are already being taken.

All employers must undertake risk assessments and, where five or more employees are employed, the assessment must be recorded in writing.

risk avoidance (1A1, 1D1, 2A1) This strategy involves a conscious decision on the part of an employer to avoid completely a particular risk by, for instance, discontinuing or modifying the activities or operations that created the risk. An example might be the replacement of manual handling operations by a mechanical handling system.

risk control (1A1, 1D1, 2A1) This is one of the outcomes of the risk management process. Risk control may be through risk avoidance, risk retention, risk transfer or risk reduction. See individual entries.

Risk Management (1A1, 1D1, 2A1) Risk management aims at producing savings in insurance premiums by first defining and then minimizing areas of industrial and other risk. It seeks not to discredit insurance arrangements but to promote the concept of insuring only what is necessary in terms of risk. On this basis, the manageable risks are identified, measured and either eliminated or controlled, and the financing of the remaining or residual risks, normally by insurance, takes place at a later stage.
Risk management has been variously defined as:

- the minimization of the adverse effects of pure and speculative risks within a business;
- the identification, measurement and economic control of the risks that threaten the assets and earnings of a business or other enterprise;
- the identification and evaluation of risk and the determination of the best financial solution for coping with the major and minor threats to a company's earnings and performance;
- the identification, analysis and degree of control exercised of risks which have the potential to threaten the assets or well-being of an enterprise. Control implies both the physical and financial steps which may eliminate, reduce or transfer the risk.

(National Examination Board in Occupational Safety and Health)
Risk management strategies include risk avoidance, risk retention, risk transfer and risk reduction. See individual entries.

risk phrase (1E1, 2E1) A phrase listed in the Approved Classification and Labelling guide accompanying the Chemicals (Hazard Information and Packaging for Supply) Regulations 1994.
A risk phrase is a standard phrase describing a hazard associated with a chemical, e.g. 'Heating may cause an explosion', 'Toxic in contact with the skin'.

Each phrase is designated by the letter R and a number.

risk rating (2A1) A term used in quantitative risk measurement and evaluation indicating a numerical level of risk. In the determination of a risk rating, factors such as probability (or likelihood), the severity of injury arising, the frequency of risk arising and the number of people exposed to the risk are taken into account. Factors such as Probability and Severity are measured on a scale from 1 to 10 or 1 to 5.

A number of formulae are used for rating risks:

1. Risk rating = Probability (P) × Severity (S) × Frequency (F)
   where all three factors are assessed on a scale from 1 to 10; this produces a risk rating from 1 to 1000.
2. Risk rating = Frequency (F) × Severity (S) + MPL + Probability (P)
   where Severity = the number of people at risk, Frequency = the frequency of the occurrence and MPL = maximum possible loss (based on a scale from 1 to 50).

In each case, the quantified risk ratings enable an employer to identify priorities for future action.

risk reduction (1A1, 2A1, 1D1, 2A3) Risk reduction, as part of the risk management process, implies the implementation within an organization of some form of loss control programme directed at protecting the organization's assets (manpower, machinery, materials and money) from wastage caused by accidental loss.

Risk reduction strategies operate in two stages:

- collection of data on as many loss-producing incidents as possible and the installation of a programme of remedial action;
- the collation of all areas where losses arise from loss-producing incidents, e.g. death, major injury, property damage, and the formulation of strategies directed at reducing these losses.

risk retention (1A1, 2A1, 1D1) A risk management strategy whereby risk is retained within an organization and any consequent loss is financed by the organization. There are two features of risk retention:

- *Risk retention with knowledge*. In this case, a conscious decision is made to meet any resulting loss from within an organization's resources. Decisions on which risks that should be retained can only be made after all the risks have been identified, measured and evaluated.
- *Risk retention without knowledge*. This generally arises from a lack of knowledge of the existence of a risk or an omission to insure against that risk. Situations where risks have not been identified and evaluated can result in this form of risk retention.

risk threshold This is the level of risk at which people are prepared to operate in terms of physical risk taking (e.g. crossing the road in heavy traffic), financial risk taking (e.g. playing the Stock Market, in gambling situations) and various forms of social risk taking (e.g. behaving in such a way that could have subsequent adverse consequences).

risk transfer (1A1, 2A1, 1D1) Risk transfer implies the legal assignments of the costs of certain potential losses from one party to another, e.g. from an organization to an insurance company.

role A pattern of behaviour that a person in a particular social status (e.g. doctor, teacher, police officer) is expected to exhibit.

role ambiguity One of the elements of role theory, role ambiguity arises where the role holder has insufficient information to adequately perform his role in the organization, or where the information provided is open to more than one interpretation.

role conflict This arises where members of an organization, who exchange information with the role holder, have different expectations of his role, and each may exert pressure on the role holder. Role conflict can result in the 'Servant of Two Masters' situation, commonly a cause of stress for the role holder.

role overload/underload Role overload results from a combination of role ambiguity and role conflict, and results from too much work to complete in the time available. Role underload, on the other hand, arises where the role holder does not have enough work, resulting in reduced job satisfaction, boredom and a feeling of isolation within the organization.

Both role overload and underload are significant causes of stress at work.

role play/simulation A technique used in simulation of real situations, designed to promote understanding of the problems involved.

An active training technique directed at increasing trainee involvement in the learning process by introducing a realistic element into instruction. Role play presents the trainees with a situation or scenario they must resolve by acting out the roles of those involved with a view to resolving the situation.

Role Theory Role theory is one of the theories concerned with stress at work. Role theory perceives large organizations as comprising systems of interlocking roles, in terms of what people do and what other people expect of them.

Stress can arise as a result of role ambiguity, role conflict and role overload/underload. See individual entries.

roofwork (1C2) One of the principal causes of fatal and major injuries in construction activities is that of people falling from or through both pitched and flat roofs. The following factors must be considered prior to and during roofwork:

- the provision of roof edge barriers (flat roofs) and working platforms;
- consideration of wind strength prior to commencing work;
- specific safety requirements when working on fragile roofs (e.g. asbestos cement panels), together with the provision of safety signs;

- safe access to the roof, with ladders placed at the correct angle and secured at their upper resting place;
- safe working practices, including the use of crawl boards and, where appropriate, personal suspension equipment (e.g safety harnesses); in certain cases, safety nets may be necessary;
- safe storage of materials on roofs;
- means for raising materials to roof level (e.g. hoists and lifting tackle).

rotating vane anemometer A specific form of anemometer for measuring air velocity, incorporating a vane which acts as a propeller. The speed of rotation is controlled by the air velocity, which is calculated from the number of vane rotations in a set time, or displayed automatically.

rotation sensing device A form of overrun device which ensures that, after the power has been interrupted or cut off, the guard remains locked and closed until the device has sensed that rotation of dangerous parts has ceased. This form of device is used with high speed mixers and centrifuges.

routes of entry (1E1, 2E1) There are three primary routes of entry of hazardous substances into the body, namely by inhalation, pervasion and ingestion. Secondary routes of entry include injection, inoculation and implantation. See individual entries.

S

safe When a vessel is properly installed, maintained and used for the purpose for which it is intended, there is no risk (apart from one reduced to a minimum) of its being the cause or occasion of death, injury or damage to property (including domestic animals).
[Simple Pressure Vessels (Safety) Regulations 1991]

safe condition sign A sign indicated by a green square or rectangle with symbols and lettering in white, e.g. FIRE EXIT.
(Safety Signs Regulations 1980)

safeguard A guard or device designed to protect persons from danger.
(BS 5304: *Safeguarding of Machinery*)

safe place of work (1C1) Section 2(2)(d) of the HSWA employers must ensure the provision and maintenance of a place of work that is, so far as is reasonably practicable, safe and without risk to health, including safe means of access and egress.
More extensive and absolute duties for ensuring a safe place of work are placed on employers under the Workplace (Health, Safety and Welfare) Regulations 1992.
The following aspects must be considered in ensuring a safe place of work in accordance with the Regulations:

- maintenance procedures for the workplace, equipment, devices and systems;
- effective and suitable provisions for ventilation;
- the provision and maintenance of a reasonable temperature;
- the provision and maintenance of suitable and sufficient lighting;
- ensuring the cleanliness of the workplace, furniture, furnishings and fittings;
- provision of adequate facilities for the storage of waste;
- prevention of overcrowding in the workplace;
- provision of suitable workstations and seating;
- maintenance of safe floors and traffic routes;
- measures to prevent falls of people and of falling objects;
- installation of safe windows, doors and gates;
- ensuring the safety of windows, skylights and ventilators;
- ensuring windows and skylights can be cleaned safely;
- organization and control of traffic routes;
- safe doors and gates;
- safe escalators and moving walkways.

Duties under the Workplace (Health, Safety and Welfare) Regulations 1992 should be read in conjunction with the more general duties on employers under the Management of Health and Safety at Work Regulations 1999 with particular reference to risk assessment, management systems, the appointment of competent persons and emergency procedures.

safe system of work (1A3) The integration of people, machinery and materials in a safe environment and workplace to produce and maintain an acceptable standard of safety.

Requirements for a safe system of work include:

- a safe layout with adequate space;
- safe means of access to and egress from the working area;
- a correct sequence of operations;
- analysis of tasks, using techniques such as job safety analysis;
- identification of safe procedures, both routine and emergency;
- a safe and healthy working environment in terms of temperature, lighting, ventilation and humidity, noise and vibration control, and hazardous airborne contamination;
- the provision of information, instruction, training and supervision for employees operating the system of work.

safety (1A1) Freedom from danger.

safety audit The systematic measurement and validation of an organization's management of its health and safety programme against a series of specific and attainable standards.

(Royal Society for the Prevention of Accidents)

A safety audit subjects each area of an organization's activities to a systematic critical examination with the principal objective of minimizing loss. It is an ongoing process aimed at ensuring effective health and safety management.

Safety Certificate A certificate issued by a local authority in cases where a sports ground provides covered accommodation for 500 or more spectators, i.e. a regulated stand. Safety certificates may further be required for stands accommodating smaller numbers of spectators.

(Fire Safety and Safety of Places of Sport Act 1987)

safety clearance In the case of simple pressure vessels, a vessel in Category A is deemed to have safety clearance once an approved body has issued an EC verification certificate or an EC certificate of conformity for such a vessel.

Category A vessels are graded on the basis of the product of the maximum working pressure in bars and their capacity in litres, thus:

- A.1 – 3000–10 000 bar litres;
- A.2 – 200–3000 bar litres;
- A.3 – 50–200 bar litres.

[Simple Pressure Vessels (Safety) Regulations 1991]

safety culture Both the HSE and CBI have provided guidance on this issue. The main principles involved, which involve the establishment of a safety culture, accepted and observed generally, are:

- the acceptance of responsibility at and from the top, exercised through a clear chain of command, seen to be actual and felt throughout the organization;
- a conviction that high standards are achievable through proper management;
- setting and monitoring of relevant objectives/targets, based upon satisfactory internal information systems;
- systematic identification and assessment of hazards and the devising and exercise of preventive systems which are subject to audit and review; in such approaches, particular attention is given to the investigation of error;
- immediate rectification of deficiencies; and
- promotion and reward of enthusiasm and good results.

(Rimington, 1989)

A company wishing to improve its performance will need to judge its existing practices against a number of features essential to a sound safety culture, namely:

- leadership and commitment from the top which is genuine and visible – this is the most important feature;
- acceptance that it is a long-term strategy which requires sustained effort and interest;
- a policy statement of high expectations and conveying a sense of optimism about what is possible supported by adequate codes of practice and safety standards;
- health and safety should be treated as other corporate aims, and adequately resourced;
- it must be a line management responsibility;
- 'ownership' of health and safety must permeate at all levels of the work force – this involves employee involvement, training and communication;
- realistic and achievable targets should be set and performance measured against them;
- incidents should be thoroughly investigated;
- consistency of behaviour against agreed standards should be achieved by auditing and good safety behaviour should be a condition of employment;
- deficiencies revealed by an investigation or audit should be remedied promptly;
- management must receive adequate and up-to-date information to be able to assess performance.

(Excerpt from *Developing a Safety Culture*, CBI, 1991)

safety data sheet Under the Chemicals (Hazard Information and Packaging for Supply) Regulations 1994 (CHIP 2), the following information must be incorporated by the supplier in a safety data sheet supplied with a substance or preparation which is dangerous for supply:

- identification of the substance or preparation;
- composition/information on ingredients;
- hazards identification;
- first aid measures;
- fire fighting measures;
- accidental release measures;
- handling and storage requirements;
- exposure controls, together with personal protection;
- physical and chemical properties;
- stability and reactivity;
- toxicological information;
- ecological information;
- disposal considerations;
- transport information;
- regulatory information;
- other information deemed appropriate.

'Substances' are chemical elements and their compounds in the natural state or obtained by any production process, including any additive necessary to preserve the stability of the product and any impurity deriving from the process used, but excluding any solvent which may be separated without affecting the stability of the substance or changing its composition.
'Preparations' are defined as mixtures or solutions of two or more substances.

safety device (1D2) A protective appliance, other than a guard, which eliminates or reduces danger before access to a danger point or area can be achieved. (BS 5304: *Safeguarding of Machinery*)
Safety devices take the form of trip devices, pressure-sensitive mat, ultrasonic device, two-hand control device, overrun device and mechanical restraint device. See individual entries.

safety factor (2D4) See ***Factor of Safety***.

safety incentive scheme A form of planned motivation, the main objectives being that of providing motivation to people by identifying the targets which can be rewarded if achieved, and making the rewards meaningful and desirable to the people involved.

safety inspection A scheduled or unscheduled inspection of a workplace or premises, or part of same, to examine current levels of safety, working practices and compliance with the relevant statutory provisions at a particular point in time.

safety needs According to Maslow, the needs of the individual for security, stability and order.

Safety of Life at Sea (SOLAS) The International Convention for the Safety of Life at Sea (SOLAS) incorporates regulations and guidance, including those related to fire safety, and adopted by the majority of maritime nations.
SOLAS covers a wide range of topics directly concerned with fire prevention

and protection, including fire detection and extinction, human element issues, fire fighting systems, wet pipe sprinkler systems, marked means of escape on ships, structural fire protection on ships, and maintenance and inspection procedures.

safety phrase (1E1, 2E1) This means, in relation to a substance or preparation dangerous for supply, a phrase listed in Part IV of the Approved Supply List to the Chemicals (Hazard Information and Packaging for Supply) Regulations 1994. In these regulations, specific safety phrases may be designated by the letter S followed by a distinguishing number or combination of numbers.
The safety phrase must be quoted in full on any label or safety data sheet in which the safety phrase is required to be shown.
Examples of safety phrases are ' Wear suitable protective clothing and gloves', 'Do not breathe the vapour'.

safety propaganda An important feature of communicating health and safety themes, hazards and messages to people, it may take the form of safety posters, films, demonstrations and exhibitions, directed at increasing awareness.

safety representative The Safety Representatives and Safety Committees Regulations 1977 are concerned with the appointment by recognized trade unions of safety representatives, the functions of safety representatives and the establishment and operation of safety committees.
The functions of a safety representative include:

- to investigate potential hazards and dangerous occurrences and examine the causes of accidents;
- to investigate health and safety complaints by the employees they represent;
- to make representations to the employer on matters arising from investigations;
- to make representations to the employer on general matters affecting health, safety and welfare;
- to carry out inspections;
- to represent employees in consultation with enforcement officers;
- to receive information from enforcement officers;
- to attend meetings of safety committees in their capacity as a safety representative in connection with the above functions.

safety sampling An organized system of regular random sampling, the purpose of which is to obtain a measure of safety attitudes and possible sources of accidents by the systematic recording of hazard situations observed during inspections made along predetermined routes in a workplace.

safety sign Safety signs must conform to a system with regard to colour and shape thus:

- Prohibition signs – circular with a red band enclosing a crossed out symbol on a white background, such as 'NO SMOKING'.
- Warning signs – triangular in shape with a yellow background and black borders, symbols and text, such as 'LPG – HIGHLY FLAMMABLE'.

- Mandatory signs – rectangular in shape, incorporating a blue mandatory symbol and/or text on a white background, such as 'WEAR EYE PROTECTION'.
- Safe condition signs – a green square or rectangle with symbols or lettering in white, such as 'FIRE EXIT'.

(Safety Signs Regulations 1980)
See Table S.1 and Figure S.1.

Table S.1 Safety signs

Meaning or purpose	**Safety colour**	**Examples of use**	**Contrasting colour (if required)**	**Symbol**
Stop Prohibition	Red	STOP signs; prohibition signs; identification of emergency shutdown devices	White	Black
Caution Risk of danger	Yellow	Warning signs, e.g. electric current on, harmful vapours, obstacle ahead, scaffold incomplete, asbestos	Black	Black
Mandatory Action	Blue	Obligation to wear personal protective equipment, e.g. eye protection; report damage immediately; keep out; switch off machine when not in use	White	White
Safe Condition	Green	Identification of first aid posts, safety showers, fire exits	White	White

Source: BS 5378, Part 1 (1980) *Safety Signs and Colours–Specification for Colour and Design.*

Employers must use a safety sign wherever there is a risk to health and safety that cannot be avoided or properly controlled by other means. Signs must contain a pictogram instead of relying solely on text. Fire safety signs and signals are included. FIRE EXIT signs must incorporate the 'Running Man' pictogram. Pipework containing dangerous substances must be marked, by fixing labels or signs at sampling and discharge points, together with small stores containing dangerous substances.
[Health and Safety (Safety Signs and Signals) Regulations 1996]

safety valve A safety device attached to a boiler and designed to open at a set pressure, higher than the normal working pressure but below the maximum working pressure, which is established during initial test of the boiler. A boiler

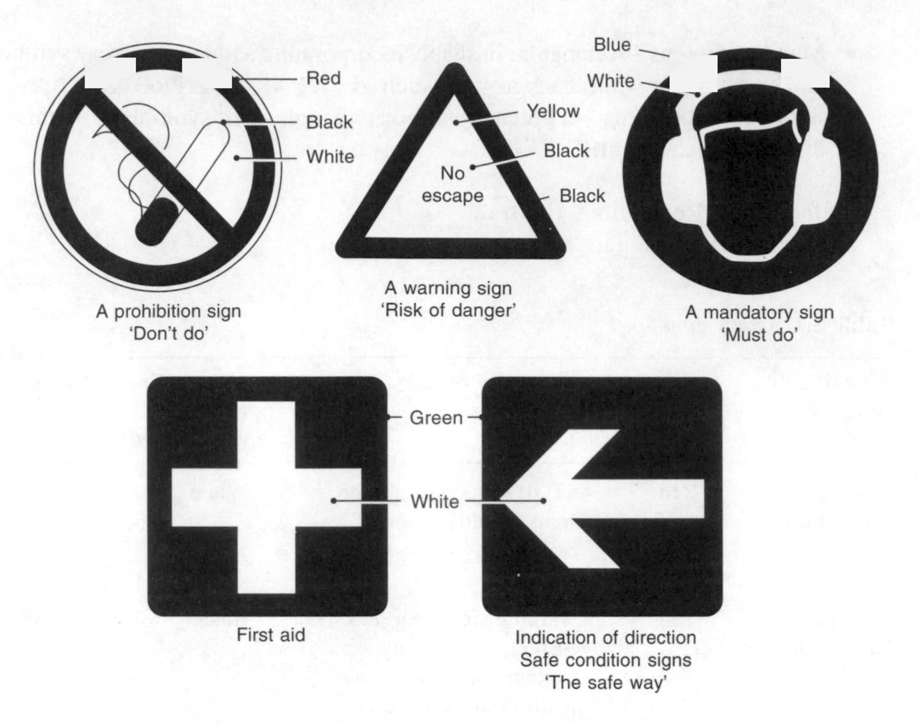

Figure S.1 Examples of classified safety signs.

should be fitted with at least two safety valves, each capable of discharging the total peak evaporation of the boiler.

Safety valves are of three types – dead weight, lever arm or steelyard and spring-loaded types.

scaffold (1C2, 2C4) A temporary form of access to structures using tubular steel members. Scaffolds take a number of forms:

- putlog or tied-in scaffolds;
- independent free-standing scaffolds;
- birdcage scaffolds.

scuffing A form of wear in moving metal surfaces arising from a failure to lubricate these surfaces. When lubrication breaks down, friction causes small heat welds between the surfaces to strip parts of the surfaces away, resulting in further wear due to these irregular surfaces.

sealed source (radiation) A source of radiation contained in such a way that the radioactive material cannot be released. The source of radiation may be a piece of radioactive material sealed in a container or held in another material which is not radioactive.

second action level A daily personal noise exposure of 90 dB(A).
(Noise at Work Regulations 1989)

secondary cutaneous sensitizer These are substances which may not necessarily bring about changes in the skin on first contact, but which produce a specific sensitization of the skin. Where further contact with the substance occurs, dermatitis will develop at the site of the second or subsequent contacts. Skin sensitizers include rubber, certain plants, nickel and many chemical substances.

secondary monitoring An area of occupational health practice directed at controlling health hazards which have already been recognized, e.g. monitoring for occupational deafness by audiometry.

secondary reinforcement The reinforcing effect of a stimulus that has been paired with a primary reinforcement.

segregation (1E3) This is a strategy aimed at controlling the hazards arising from toxic substances and certain physical hazards, such as noise and radiation. Segregation may take a number of forms:

- segregation by distance (separation);
- segregation by age;
- segregation by time;
- segregation by sex.

Self-Actualization (Self-Actualizing Man) The ultimate stage in Maslow's hierarchy of motivation, where the individual achieves his true potential. It is synonymous with personal development, self-fulfilment, creativity, happiness and growth.

self-adjusting automatic guard A guard which prevents accidental access of a person to a danger point or area but allows the access of a workpiece which itself acts as part of the guard, the guard automatically returning to its closed position when the operation is completed.
(BS 5304: *Safeguarding of Machinery*)

self-contained breathing apparatus A form of breathing apparatus which may be of the open or closed circuit type.
In the first case, air is supplied by a lung-governed demand valve or pressure reducer connected to a full face piece via a hose supply. The hose is connected to its own compressed air or oxygen supply, which is carried by the user in a harness.
With the closed circuit type, a purifier is incorporated to absorb exhaled carbon dioxide. The purified air is fed back to the respirator after mixing with pure oxygen.

self-defensive function According to Katz, this is one of the four functions of attitude. An individual needs to defend his self-image, which may be from an external viewpoint in terms of how he or she is perceived by others, and an internal viewpoint, in terms of his or her inner impulses and personal knowledge of what he or she really is like.

self-image This is the image that a person wishes to project to those with whom he or she comes into contact. A person may wish to be perceived as, for instance, fair-minded, hard-minded, friendly, difficult to fool, strict, easy to get on with, etc.

self-regulation (2B2) The philosophy, outlined in the Robens Report (1970–1972), whereby organizations regulate their health and safety activities within certain general parameters specified in the Health and Safety at Work etc. Act 1974.

sensitization (1E1, 2E4) An increased reacton on the second or subsequent exposure to a substance due to an immunological mechanism.
Certain substances and preparations may have an effect on the persons exposed. When such substances are inhaled or penetrate the skin, these substances are capable of eliciting a reaction by hypersensitization, such that on further exposure to the substance or preparation, characteristic adverse effects are produced. These substances and preparations must be appropriately labelled in accordance with the Chemicals (Hazard Information and Packaging for Supply) Regulations 1994. Hypersensitization principally affects the lungs and skin, with the result that even the smallest exposure to the substance or preparation will produce some form of response, such as asthma and dermatitis.

sensitizing Substances and preparations which, if they are inhaled or if they penetrate the skin, are capable of eliciting a reaction by hypersensitization such that, on further exposure to the substance or preparation, characteristic adverse effects are produced.
[Chemicals (Hazard Information and Packaging for Supply) Regulations 1994]

sequential operation A method of operating machinery employing electrical, mechanical, hydraulic or pneumatic means to set certain mechanisms in motion automatically, in sequence, following the closure of a guard.
(BS 5304: *Safeguarding of Machinery*)

series manufacture Where more than one pressure vessel of the same type is manufactured during a given period by the same continuous manufacturing process, that is in accordance with a common design.
[Simple Pressure Vessels (Safety) Regulations 1991]

shearing trap A form of machinery hazard associated with two pieces of metal shearing together and creating a cutting effect, e.g. guillotines, hand shears. See Figure S.2.

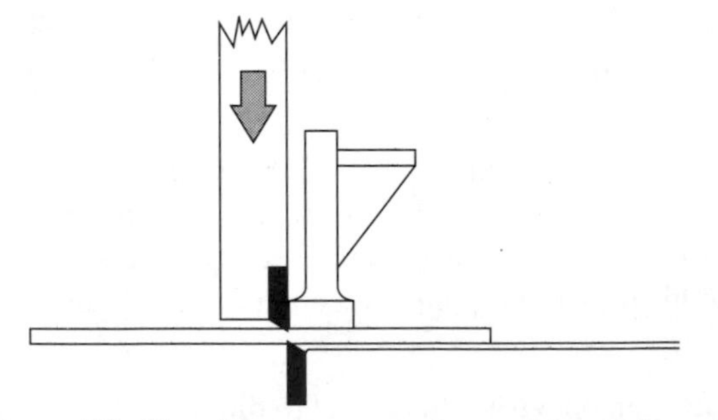

Figure S.2 Shearing trap – a moving part traversing a fixed part.
Source: BS 5304 (1988) Safeguarding of Machinery.

shear strain (2D4) This is expressed by the amount of deformation in one part of a body compared with another part that has not moved.

$$\text{Shear strain} = \frac{\text{Deformation}}{\text{Distance between deformed layer and static layer}}$$

As strain is a ratio, there are no units.

shear stress (2D4) A form of stress to which materials may be subjected when forces act either in opposite directions or towards each other, but not in the same line of action.

$$\text{Shear stress} = \frac{\text{Applied force}}{\text{Cross-section of area in shear}}$$

shielding A form of radiological protection involving the absorption of radiation energy by the interaction with a dense medium, such as lead and concrete.

short circuit This arises where a current finds a path from live to return other than through the apparatus, resulting in high current flow, heating of conductors to white heat and arcing.

short-term exposure limit (1E2) See ***Occupational Exposure Limits***.

shut down maintenance (2A3) Maintenance of a workplace, plant and equipment undertaken when the workplace is closed, e.g. during holidays and weekend periods.

simple asphyxiant An inhaled substance which causes oxygen deficiency in the lung, e.g. carbon dioxide, hydrogen, nitrogen.

simple pressure vessel A vessel made of certain types of steel or aluminium that is intended to contain air or nitrogen under pressure and manufactured in series. [Simple Pressure Vessels (Safety) Regulations 1991]

Single Burning Item (SBI) fire test This is a standard for assessing the fire resistance of building products throughout EU countries. Products which meet this standard are awarded a CE mark. The test classifies materials in one of six groups (A–F), with Class F having the lowest performance.

Single European Act A statute which identified the need to eliminate technical barriers to trade, one of which was the wide variation in legal standards relating to health and safety throughout the European Community. The Act introduced the 'new approach' to technical harmonization, the development of the 'essential safety requirements' and of harmonization Directives that lay down the essential safety requirements.

single phase system A single phase AC electrical system, normally operating at 240 volts, in which the circuit comprises two conductors ('live' and 'neutral'), generally with an additional conductor for 'earth'.

smoke A product of incomplete combustion, mainly of organic material. It may contain ash, soot and grit.

smoke detector A device which detects the present of smoke in a building. There are three principal types, namely ionizing detectors, light scatter detectors and light obscuration detectors.

smothering A means of fire spread control which can be achieved by allowing a fire to consume the oxygen present, but preventing the inward flow of more oxygen, or adding an inert gas to the burning mixture.

social adjustment function According to Katz, this is one of the four functions of attitude. It is concerned with how people relate and adjust to the influences of, for instance, friends, teachers, parents and superiors. The resulting behaviour is based, to some extent, on a philosophy of 'maximum reward, minimum punishment'.

sociocusis Noise-induced hearing loss; occupational deafness.

soda acid extinguisher The original form of water-containing fire extinguisher which is gradually being replaced by other types of appliance. With this type of extinguisher, gas is generated in the cylinder when the acid phial is broken by the plunger, water being expelled from a discharge tube.

software ergonomics With the majority of display screen equipment, the presentation of information is controlled by the software. Software design is, therefore, an important element of task design. Procedures should be established within an organization to ensure appropriate design, selection and modification of software. Software should be suitable for the task, easy to use and adaptable for users, provide feedback on the performance of the system and operate at a pace and with a format that is comfortable for users.

Under the Health and Safety (Display Screen Equipment) Regulations 1992, the above factors should be considered by an employer in a display screen equipment workstation risk analysis.

sole plate In tubular steel scaffolding, a strong timber plank or section, wider than a standard base plate, and long enough to be positioned under at least two base plates. The purpose of a sole plate is to spread to the load imposed on standards, particularly where the ground-bearing capacity may be poor or uneven.

somatic effect An effect on a person who has been exposed to radiation, which has caused direct damage to body cells.

Acute effects arising from a large dose of radiation in a short period of time include nausea, vomiting and depletion of white blood cells, followed by death within approximately 1 month. Other effects include erythema, loss of hair, rapid weight loss, skin pigmentation changes and skin lesions.

Chronic effects include cataracts of the eye and cancer.

somatic system The part of the nervous system serving the sense organs and the skeletal muscles.

sone This is a linear unit of loudness on a scale designed to give readings approximately proportional to loudness. This scale is precisely defined by its relation to the phon scale.

sound (2E2) Any pressure variation in air, water or some other medium that the human ear can detect.

sound intensity/sound power (2E2) The particular power of a sound or the level of sound energy with which it confronts the ear. Sound intensity describes the rate of flow of sound energy.

sound pressure (2E2) The passage of sound energy through air is accompanied by fluctuations in atmospheric pressure. These fluctuations can be measured and related to the amount of sound energy that is flowing. Sound pressure is a measure of the magnitude of the air pressure variations of fluctuations which make up sound. The root mean square value of the pressure variations is used and expressed in decibels (dB).

special damages Damages awarded in a civil action which relate to losses incurred after the hearing of the action. These losses may be actual loss of earnings and probable loss of future earnings following an accident.

specific safety certificate A safety certificate for a specific sporting activity or occasion.
(Fire Safety and Safety of Places of Sport Act 1987)

spillage containment (1C6) All measures taken to contain spillages by using absorbent materials (e.g. sand, sawdust) and other techniques, such as the use of movable bunds and enclosures.

spontaneous combustion (spontaneous ignition) A characteristic of certain substances where the substance bursts into flames without apparent source of ignition (e.g. haystacks, oil-soaked lagging). Spontaneous ignition is brought about by a process of oxidation of the substance, the heat generated from this process eventually providing an ignition source.

spontaneous ignition temperature The lowest temperature at which a substance will ignite spontaneously.

spreader bar (lifting beam) A device for ensuring safe lifting of loads which incorporates a steel beam, the means of attaching the beam to a lifting machine and the means of fixing the beam to the load to be raised. This device enables a particular load to be lifted in a particular way, often to prevent horizontal stressing of eyebolts. See Figure S.3.

sprinkler system A permanent form of installation that provides an automatic means of detecting and extinguishing or controlling a fire in its early stages.
A typical sprinkler system incorporates:

- an overhead pipe system incorporating sprinkler heads at suitable intervals;
- water supplied from a head tank or main.

The sprinkler heads act as valves which are preset to open at a given temperature and spray water on to a fire. Spacing of sprinkler heads and the quantity of water delivered depend upon the classification of fire risk. See Figure S.4.

Figure S.3 Typical lifting beam (spreader bar).

stain tube detector (1E2) A device used in short-term air sampling (grab sampling or snap sampling), which gives an immediate indication of the concentration in air of a particular airborne contaminant. The detector incorporates a graduated glass stain tube, sealed at both ends, and filled with porous granules of an inert material, such as silica gel. The granules are impregnated with a chemical agent which changes colour in the presence of the contaminating agent, such as a gas. The stain tube detector is used in conjunction with a device, such as hand-operated bellows or piston, which draws air through the tube, the length of the stain (change in colour) indicating the degree of contamination by the particular contaminant. Stain tubes are commonly available for the detection of over 1000 airborne contaminants and provide an early indication as to whether more sophisticated air sampling techniques may be necessary.

standard In tubular steel scaffolding, a vertical tube used as a support for transferring the load to the ground or to a base plate.

stare decisis A Latin term meaning 'let the decision stand', implying the need to stick to the decisions of past cases. *Stare decisis* is the basis for the doctrine of binding precedent.

starvation A form of fire spread control which can be achieved by taking the fuel away from a fire, taking the fire away from the fuel and reducing the quantity or bulk of the fuel available.

statement of claim A statement, outlining the allegations against a defendant, and previously delivered to the defendant prior to a civil action.

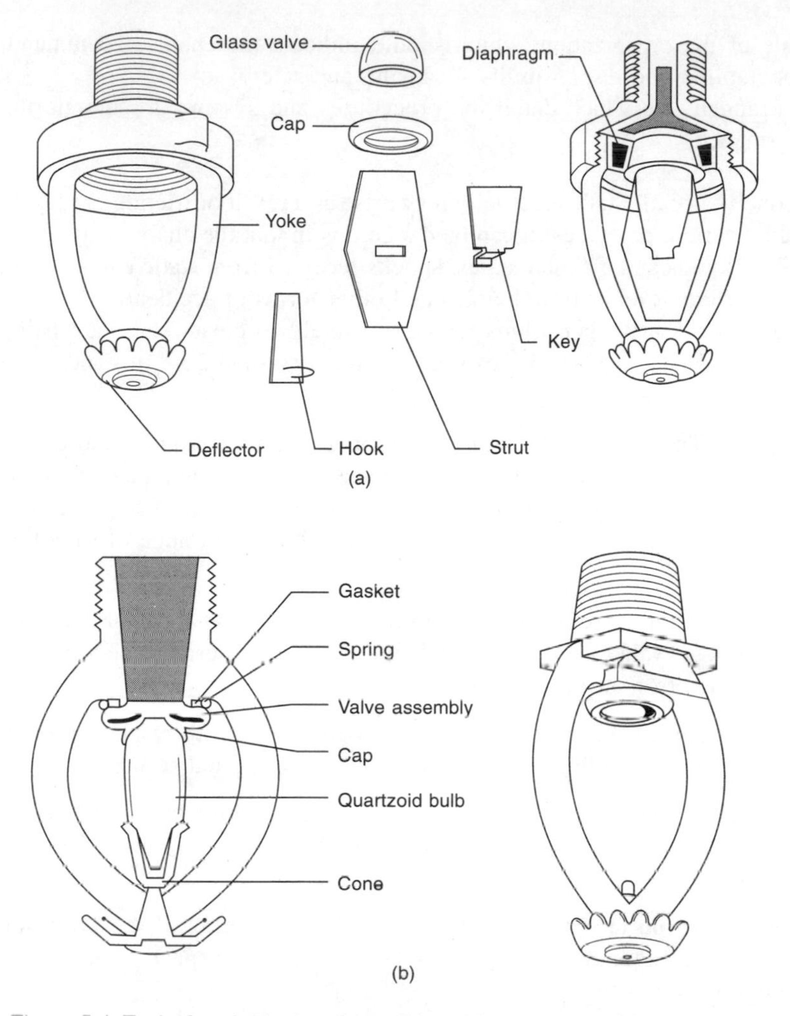

Figure S.4 Typical sprinkler heads: fusible solder type (a) and bulb type (b). Source: Manual of Firemanship. London: Home Office.

Statement of Health and Safety Policy (1B2) There is a general duty under section 2(3) of the HSWA for an employer to prepare, and as often as may be necessary, revise a written statement of his or her general policy with respect to the health and safety at work of his employees and the organization and arrangements for the time being in force for carrying out that policy.
The statement and any revision of same must be brought to the notice of employees. The statement need not be written where less than five employees are employed. The principal features of a statement of health and safety policy are:

- a statement of intent which outlines the organization's overall philosophy in relation to the management of health and safety, including objectives for ensuring legal compliance;

- details of the organization, which should indicate the chain of command, accountability and responsibility for health and safety;
- the arrangements which detail the procedures and systems for monitoring performance.

static electricity (2C2) Static electricity may arise as a result of friction or may be induced by certain processes. It can be dangerous in that the charge may ignite flammable liquids and, in some cases, shocks received from static can result in falls on the same level or from heights and other forms of accident.
Earthing is necessary when fluids which are likely to create static are being transferred and all pipes and hoses used for transfer should be of conducting material.

static sampling The process of sampling airborne contamination by the use of fixed sampling devices located in the workplace which sample on a continuous basis or for a specified period of time. Static sampling operates using mains or battery operated pumps, passing the sampled air through a range of sampling devices.

statistics A collection of techniques used in the quantitative analysis of data, and used to facilitate evaluation of the data. Numbers used to describe distributions and to estimate errors of measurement.

status needs The needs of an individual to achieve status with respect to other people in a group, including the need for power, prestige and security.

statute (1B1, 1B2) Statutes (Acts of Parliament) are the written laws produced as a result of the parliamentary process. They supersede all other forms of law, and only Parliament can make, modify, revoke or amend statutes.

statutory examination (1D1) With reference to Regulation 32 of the Provision and Use of Work Equipment Regulations 1998, thorough inspection in relation to a thorough examination:

- means a thorough examination by a competent person;
- includes testing the nature and extent of which are appropriate for the purpose described in the regulation.

statutory inspection (1C2) With reference to Regulation 6 of the Provision and Use of Work Equipment Regulations 1998:

- such visual or more rigorous inspection by a competent person as is appropriate for the purpose described in the regulation;
- where it is appropriate to carry out testing for the purpose, includes testing the nature and extent of which are appropriate for the purpose.

Statutory Instrument A statute confers powers on a Minister or Secretary of State to make statutory instruments. Statutory instruments, in most cases, take the form of Regulations (delegated or subordinate legislation) and provide detailed requirements on specific matters, such as the workplace, work equipment and manual handling operations.

statutory nuisance Under section 79 of the EPA, the following are defined as statutory nuisances:

- any premises in such a state as to be prejudicial to health or a nuisance;
- smoke emitted from premises so as to be prejudicial to health or a nuisance;
- fumes or gases emitted from premises so as to be prejudicial to health or a nuisance;
- any dust, steam, smell or other effluvia arising on industrial, trade or business premises and being prejudicial to health of a nuisance;
- any accumulation or deposit which is prejudicial to health or a nuisance;
- any animal kept in such a place or manner as to be prejudicial to health or a nuisance;
- noise emitted from premises so as to be prejudicial to health or a nuisance;
- any other matter declared by any enactment to be a statutory nuisance.

steam boiler Any closed vessel in which, for any purpose, steam is generated under pressure greater than atmospheric and includes any economizer used to heat water being fed to any such vessel, and any superheater used for heating steam.
(Factories Act 1961)

steam container Any vessel, other than a steam pipe or coil, constructed with a permanent outlet to the atmosphere or into a space where pressure does not exceed atmospheric pressure, and through which steam is passed at atmospheric pressure, or at approximately that pressure, for the purpose of heating, boiling, drying, evaporating or other similar purpose.
(Factories Act 1961)

steam receiver Any vessel or apparatus, other than a steam boiler, steam container, steam pipe or coil, or part of a prime mover, used for containing steam under pressure greater than atmospheric pressure.
(Factories Act 1961)

stimulus Any object, energy or energy change in the physical environment that excites a sense organ.

stop valve A valve located between a boiler and steam pipe or outlet, used to control the flow of steam from the boiler.

stored pressure appliance There are two types of stored pressure appliance:

- A water type stored pressure appliance contains carbon dioxide under pressure, the water being released when the trigger is pulled.
- A foam type stored pressure appliance contains foam concentrate and is pressurized to 10 bar with air or nitrogen. Operation of the trigger valve allows the pressure to expel the foam concentrate through the exit pipe, where foam is generated at the end of the hose.

strain (2D4) Where a material is subjected to some form of elongation, compression or shortening per unit length it is subjected to strain.
Strain is a ratio of the extension/compression of the material to its original length.

$$\text{Strain} = \frac{\text{Elongation (or shortening)}}{\text{Original length}}$$

As strain is a ratio, there are no units.

stress (2D4) Where a material is subjected to tension or compression, it is under stress.

$$\text{Stress} = \frac{\text{Load}}{\text{Cross-sectional area}}$$

Stress is measured in newtons/square metre.

stress corrosion (2D4) This is a common cause of failure in metal components, whereby fracture results from the combined influence of corrosion with tensile stress.

stress (psychological) (2E6, 2E7) Stress can be defined in a number of ways:

- the common response to attack (Selye, 1936);
- the common response to environmental change;
- a feeling of sustained anxiety which, over a period of time, leads to disease;
- the non-specific response of the body to any demands made upon it;
- a psychological response which follows failure to cope with problems.

Causes of stress can be classified as environmental stressors, occupational stressors and social stressors.
Environmental stressors include extremes of temperature, poor lighting conditions, inadequate ventilation, noise, vibration and airborne contaminants.
Social stressors are associated with family life, marital relationships, bereavement, etc. – the everyday problems of coping.
Occupational stressors are associated with work and include:

- too heavy or too light a workload;
- a job which is too difficult or too easy;
- working excessive hours;
- conflicting job demands – the 'servant of two masters' situation;
- too much or too little responsibility;
- poor working relationships;
- incompetent superiors;
- lack of participation in decision making;
- middle age vulnerability associated with reduced career prospects or the need to change career;
- overpromotion or underpromotion;

- interaction between work and family commitments;
- deficiencies in interpersonal skills.

See Figure S.5.

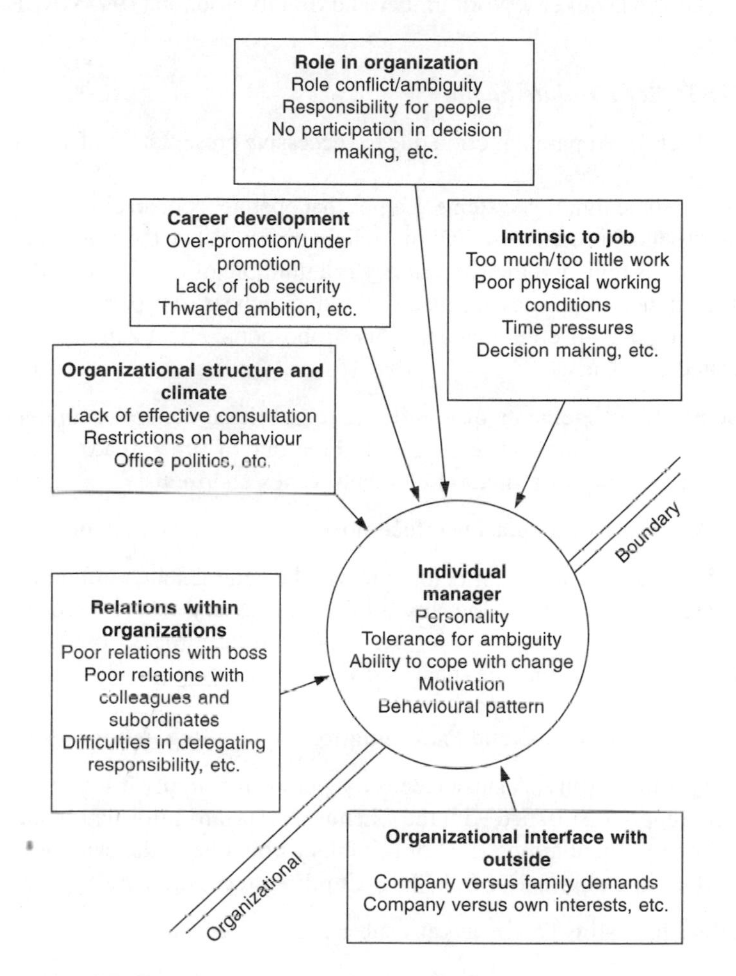

Figure S.5 Sources of managerial stress. Source: *Cooper and Marshall (1978).*

Signs and symptoms of stress are, in many cases, specific to an individual and can include headaches, inability to sleep, overeating, constipation, lower back pain, allergies, nervousness, nightmares, high blood pressure, alcohol abuse, indigestion, dermatitis, menstrual distress, nausea, irritability, loss of appetite, asthma attacks, heart palpitations etc.; insomnia is a classic indicator of stress. Stress-induced injury has been the subject of a number of civil claims recently, where:

- a ticket collector was awarded £375 000 damages against his employer, London

Underground Ltd, for the condition known as post-traumatic stress disorder, suffered as a result of the King's Cross fire;
- the High Court held an employer liable for the psychiatric damage suffered by an employee after they failed to take reasonable steps to avert a second nervous breakdown [Walker v. Northumberland County Council (1995) IRLR 35].

strict liability (2B3) See ***absolute liability***.

stroboscopic effect (2E3) Apparent motion due to successive presentation of visual stimuli.
This is a feature of some lighting systems. Lamps that operate from an alternating current electricity supply produce oscillations in light output. When the magnitude of these oscillations is high and their frequency is a multiple of the frequency of movement of machinery, that moving machinery may appear to a person to be stationary or moving in a different manner. This stroboscopic effect can result in machinery-related accidents.

structural elements Those elements of a building (e.g. bricks, concrete, timber) which are significant in terms of fire control and the rate of fire spread.
The relative combustibility of structural elements varies enormously.

sub-acute effect A reduced form of acute effect of a substance on the human body.

substance Chemical elements and their compounds in the natural state or obtained by any production process, including any additive necessary to preserve the stability of the product and any impurity deriving from the process used, but excluding any solvent which may be separated without affecting the stability of the substance or changing its composition.
[Chemicals (Hazard Information and Packaging for Supply) Regulations 1994]

substance dangerous for supply A substance is dangerous for supply if it is listed in the Approved Supply List issued with the Chemicals (Hazard Information and Packaging for Supply) Regulations 1994 or if it falls within one of the categories of danger specified in Column 1 of Schedule 1 to these regulations.

substance hazardous to health This means any substance (including any preparation) which is:

(a) a substance which is listed in Part 1 of the Approved Supply List as dangerous for supply within the meaning of the Chemicals (Hazard Information and Packaging for Supply) Regulations 1994 and for which an indication of danger specified for the substance in Part V of that list is very toxic, toxic, harmful, corrosive or irritant;
(b) a substance for which the HSC has approved a maximum exposure limit or an occupational exposure standard;
(c) a biological agent;
(d) a substance, not being a substance mentioned in the sub-paragraphs above, which creates a hazard to the health of any person which is comparable with the hazards created by substances mentioned in those sub-paragraphs;

(e) dust of any kind, except dust which is a substance within the first two paragraphs above, when present at concentrations in air equal to or greater than:
   - 10 mg/m^3, as a time-weighted average over an 8-hour period of total inhalable dust; or
   - 4 mg/m^3, as a time-weighted average over an 8-hour period of respirable dust.

(Control of Substances Hazardous to Health Regulations 1999)

substitution A prevention strategy whereby a less hazardous substance, process or work activity is substituted for a more dangerous one.

suitable head protection Head protection that:

- is designed to provide protection, so far as is reasonably practicable, against foreseeable risks of injury to the head to which the wearer may be exposed;
- after any necessary adjustment, fits the wearer;
- is suitable with regard to the work or activity in which the wearer may be engaged.

[Construction (Head Protection) Regulations 1989]

summary offences Criminal offences which can only be dealt with in a court of summary jurisdiction, e.g. a Magistrates' Court. Such offences are mainly of a minor nature.

summons A written order, signed by a magistrate, requiring a named individual to appear before a particular court at a specified time on a specified date to answer a particular charge or charges.

superego In psychoanalytical theory, that which restrains the activity of the ego and the id. The superego corresponds closely to conscience, and keeps a person working towards ideals developed during childhood.

supplier A person who supplies a substance or preparation dangerous for supply in the course of, or for use at, work by way of sale, or offer for sale, commercial sample or transfer from any place of work (whether or not in the same ownership) and, in the case of a substance, includes the importer into Great Britain, whether or not that substance is imported from a European Member State.
[Chemicals (Hazard Information and Packaging for Supply) Regulations 1998]

Supreme Court of Judicature A collective term for the High Court, Crown Courts and the Court of Appeal.

sustainable development Making the best possible use of unavoidable waste and minimizing the risk of pollution or harm to human health arising from waste disposal or recovery.
(Environmental Protection Act 1990)

swing vane anemometer (velometer) A form of anemometer incorporating a hinged vane, which is deflected by air movement, giving a direct scale reading.

syndicate exercise An active form of training whereby a larger group can be broken into smaller groups for discussion or problem-solving exercises, with the trainer available for consultation and guidance.
This technique allows for the sharing of experiences, decision making on a group basis and ensures discipline in solving a particular problem. Objectives must be clearly established and a report back session should ensure the solution to the problem has been reached.

system Any electrical system in which all the electrical equipment is, or may be, electrically connected to a common source of electrical energy, and includes such source and such equipment.
(Electricity at Work Regulations 1989)

systematic training (1CS2) For training to be effective it must be systematic, implying the following:

- the presence of a trained and competent trainer/instructor working with suitable trainees;
- defined training objectives which are measurable by trainers and achievable by the trainees;
- a content of knowledge broken down into sequential units which can be readily assimilated;
- a content of skills analysed by elements;
- a clear and orderly training programme;
- an appropriate place in which to learn;
- suitable equipment and visual aids;
- sufficient time for trainees to attain a desired standard of knowledge and competence, with frequent testing to ensure trainees understand and know what has to be learned.

(Department of Employment)

systemic effect An effect of an exposure to a hazardous substance which affects a particular body system, such as the central nervous system or digestive system.

systems and sub-systems failure (2A2) The unintentional release of stored energy (other than from a pressure relief system) from a pressure system or sub-system.
(Pressure Systems Regulations 2000)

T

target organ and target system (2E1) Certain toxic substances have a direct or indirect effect on specific body organs (target organs) and body systems (target systems).
Target organs include the liver, lungs, bladder, brain and skin. Target systems include the respiratory system, circulatory system, lymphatic system and reproductive system.

Task Analysis (1A3) See *job (task) analysis*.

task design Inadequate or unsuitable task design can be a cause of stress amongst display screen equipment users. Task design should ensure users have:

- variety to exercise discretion, opportunities for learning and appropriate feedback;
- sufficient work to prevent overload or underload;
- have the chance to participate in the planning, design and implementation of tasks.

teagle opening An opening in the fabric of a building above ground level through which goods can be lowered to the outside or hoisted into the building. Teagle openings are a common feature of warehouses, flour mills and agricultural buildings. Teagle openings must be securely fenced and provided with a secure handhold at each side to prevent people from falling through the opening.

technical file (2D2) In the case of potentially dangerous new machinery, a file of information which a machinery supplier must send to an approved test house detailing how the machinery complies with European harmonized standards, or, if it does not comply, a sample of the machine itself for type approval.
[Supply of Machinery (Safety) Regulations 1991]

Technique for Human Error Rate Probability (THERP) A technique for predicting the potential for human error in an activity. It evaluates quantitatively the contribution of the human error component in the development of an untoward system. Special emphasis is placed on the human component in product degradation. THERP involves the concept of a basic error rate that is relatively consistent between tasks requiring similar human performance elements in different situations. Basic error rates are assessed in terms of contributions to specific systems failures. The methodology of THERP entails:

- selecting the system failure;

- identifying all behaviour elements;
- estimating the probability of human error;
- computing the probabilities as to which specific human error will produce the system failure.

Following the classification of probable errors, specific corrective actions are introduced to reduce the likelihood or error.

temporary employment A situation where an employee of one organization may be hired out to another organization to undertake a specific task. The test of whether an employee has been temporarily employed is based on the extent of control (the control test) that the temporary employer can exert in that employee's actual undertaking of the specific task involved.
[Mersey Docks & Harbour Board v. Coggins and Griffiths (Liverpool) Ltd (1974) AC1]

temporary threshold shift (temporary deafness) A short-term loss of hearing acuity (or shift) as a result of exposure to loud noise. The condition is reversible in most cases, although individual susceptibility to noise may result in some degree of permanent shift in certain people exposed.

tensile stress A form of stress in a material arising from stretching of the material by forces acting in the same line but in opposite directions.

$$\text{Tensile stress} = \frac{\text{Applied force/load}}{\text{Cross-sectional area of solid}}$$

teratogenesis The process of interference with foetal development in such a manner that an abnormal foetus is produced.

Theory of Management and Organization of Work This first theory of motivation was proposed by Taylor in 1911. Taylor said 'Man is a creature who does everything to maximize self-interest'. In other words, if the money was right, people would do virtually anything to earn money.
Taylor examined the differences between managers, who were involved in the organization, planning and supervision of work, and the rest of the workforce, who did not appear interested in these matters. They preferred to have simple tasks organized for them that they could be trained to undertake without having to make decisions.
This viewpoint, that people only went to work for money, was held by many organizations for years, but has now largely been discredited.

Theory of Unconscious Motivation This theory of accident causation, which has its roots in psychoanalytical theory, states that accidents are brought about by subconscious processes, in particular, aggression, guilt, ambition, conflict and anxiety.

Theory X This theory (along with Theory Y) of McGregor, in his book *The Human Side of Enterprise* (1960), is significant in the role of organizations and how they operate. Theory X was based on the observation of human behaviour, and examined

attitudes to work brought about by people's experiences of the conditions existing in organizations.
Theory X postulates:

- the average individual has an inherent dislike of work and will avoid it if he or she can;
- due to this dislike for work, most people have to be coerced, controlled, directed and treated with punishment in order to achieve the organization's objectives;
- the average individual prefers to be directed, wishes to avoid responsibility and wants security above all.

For some people, pay and job security mean more than advancement and satisfaction with work.

Theory Y This theory by McGregor (1960) is based on the view that people are not passive or resistant to organizational needs by nature, but have been made so by their experience of working for organizations. The motivation, the capacity for responsibility and directing behaviour towards the organization's goals are naturally present in people. Management does not put them there. Management must, therefore, recognize and develop these characteristics in workers, through better organizational conditions and methods of operation, enabling people to achieve greater job satisfaction through their own efforts.

thermal anemometer A form of anemometer, an electronic instrument which detects the change of resistance of a heated wire sensor or thermistor when it is cooled by the movement of air.

thermal comfort (1C1, 1E1) The sensation of comfort associated principally with ambient temperature, but also relative humidity and the rate of air movement.

thermoluminescent dosemeter A type of personal dosemeter which operates on the principal that when a material, such as lithium fluoride, is bombarded with ionizing radiation, it converts to an excited state. This state is reversed only on the application of heat, when the crystals return to normal, but with a measurable emission of light. A small badge containing these crystals will detect radiation, the amount of radiation being related to the amount of light produced on heating.

thorough examination In relation to a thorough examination under Regulation 9 (Thorough Examination and Inspection):

- means a through examination by a competent person;
- where it is appropriate to carry out testing for the purpose described in the paragraph, includes such testing by a competent person as is appropriate for the purpose.

(Lifting Operations and Lifting Equipment Regulations 1998)

three-phase system An AC system, normally at 415/240 volts, incorporating four conductors, three of these conductors being live and the fourth neutral. This type of system is generally used for larger items of electrical equipment and the maximum voltage of the system (415 volts) is utilized.

threshold The level of stimulus energy which must be exceeded before a response occurs.

Threshold Limit Value – Time-Weighted Average (TWA) (1E2) The threshold limit value TWA represents the time-weighted average concentration of atmospheric contamination present for a normal 8-hour day (or 40-hour week) to which nearly all workers may be repeatedly exposed, day after day, without adverse effect.

time delay interlock A form of machinery interlock system featuring a mechanical device which takes a specific period of time for a guard to be released after the power has been switched off.

tinnitus The sensation of whistling or ringing in the ears as a result of, principally, chronic exposure to noise.

torsional stress (2D4) The stress applied to materials subject to a turning effect, as with rotating shafts in machinery.

torts A tort is a civil wrong. The three principal torts are those of negligence, nuisance and trespass.
The rule of common law that everyone owes a duty to everyone else to take reasonable care so as not to cause them foreseeable injury.

Total Accident Control A stage of a Total Loss Control programme directed at the prevention of all accidents resulting in personal injury and/or property damage. Three steps towards Total Accident Control are spot checking systems, reporting by control centres and safety audits.

total inhalable dust Airborne material which is capable of entering the nose and mouth during breathing and is thereby available for deposition in the respiratory tract.
(Control of Substances Hazardous to Health Regulations 1999)

Total Loss Control A programme designed to eliminate or reduce all accidents which downgrade the system, and which result in wastage of an organization's assets.
An organization's assets are manpower, materials, machinery, manufactured goods and money (the five 'Ms').

toxic Substances and preparations which in low quantities can cause death or acute or chronic damage to health when inhaled, swallowed or absorbed via the skin. [Chemicals (Hazard Information and Packaging for Supply) Regulations 1994]

toxic for reproduction Substances and preparations which, if they are inhaled or ingested or if they penetrate the skin, may produce or increase the risk of non-heritable adverse effects in the progeny and/or impairment of male or female reproductive functions or capacity.

[Chemicals (Hazard Information and Packaging for Supply) Regulations 1994]

toxicity (2E5) The ability of a chemical molecule to produce injury once it reaches a susceptible site in or on the body.

Toxicological Assessment (2E5) The collection, assembly and evaluation of data on a potentially toxic substance and the conditions of its use, in order to determine:

- the danger to human health;
- systems for preventing or controlling the danger;
- the detection and treatment of overexposure; and
- where such information is insufficient, the need for further investigation.

The following factors should be considered in toxicological assessment:

- The name of the substance, including any synonyms.
- A physical and/or chemical description of the substance.
- Information on potential exposure situations.
- Details of occupational exposure limits.
- General toxicological aspects, such as:
  - the route of entry into the body;
  - the mode of action in or on the body;
  - signs and symptoms;
  - diagnostic tests;
  - treatment; and
  - disability potential.

toxicology (2E5) The study of the body's responses to toxic substances.

traffic route A route for pedestrian traffic, vehicles or both, and includes any stairs, fixed ladder, doorway, gateway, loading bay or ramp.
[Workplace (Health, Safety and Welfare) Regulations 1992]

training (1CS2) The systematic development of attitude, knowledge and skill patterns required by an individual to perform adequately a given task.
(Department of Employment)

training need (1CS2) A training need exists when the optimum solution to an organization's problem is through some form of training.

transfer of training This is an important factor in the psychology of learning. Transfer of training may be negative or positive.
Positive transfer takes place where more rapid learning is achieved through it being assisted by previous learning and experience in another situation.
Negative transfer, on the other hand, can arise where learning is slower because of previous learning in an other situation which confuses the trainee or retards the acquisition of skills and knowledge.

transmission machinery Every shaft, drum, pulley, system of fast and loose pulleys,

coupling, clutch, driving belt or other device by which the motion of a prime mover is transmitted to or received by any machine or appliance.
(Factories Act 1961)

transverse stress (2D4) This is created when a material is subjected to forces which cause it to bend. This may arise in the overloading of a structure, whereby the supporting beams deflect downwards. It is a combination of tension and compression produced at the same time by the action of forces on a body.

traps A form of machinery hazard which may take the form of a reciprocating trap, shearing trap or in-running nip. See individual entries.

travel This term, which is featured in the definition of 'means of escape', means:

- travel within rooms;
- travel from rooms to a stairway or final exit;
- travel within stairways and to a final exit.

trespass Trespass falls within the law of tort. It implies the intentional invasion of an individual's person, land or goods.
An action for trespass involves a civil claim for damages resulting from false imprisonment, unlawful entry on to the land of another, assault and battery.

trespasser A trespasser is defined in common law as a person who:

- goes on to premises without invitation or permission;
- although invited or permitted to be on the premises, goes to a part of the premises to which the invitation or permission does not extend;
- remains on premises after the invitation or permission to be there has expired;
- deposits goods on premises when not authorized to do so.

tribunal A legally constituted body consisting of a chairman appointed by the Lord Chancellor and two lay members, one representing employers and the other the trade unions. Members are selected from panels maintained by the Department of Employment following nominations from employers' organizations and trade unions. When all three members are sitting, the majority view prevails.

trip device A means whereby any approach by a person beyond the safe limit of working machinery causes the device to actuate and stop the machinery or reverse its motion. See Figure T.1.
(BS 5304: *Safeguarding of Machinery*)

tumour (neoplasm) A mass of cells which have undergone some fundamental and irreversible change in the physiology and structure, which leads to a continuous and unrestrained proliferation of such cells.
Benign tumours are those which grow slowly and less expansively and, unless they occur in some vital site, are generally well tolerated.
Malignant tumours grow more rapidly and will infiltrate and extend into normal tissue and body structures. In many cases, when left untreated, they interfere with general health and may cause death.

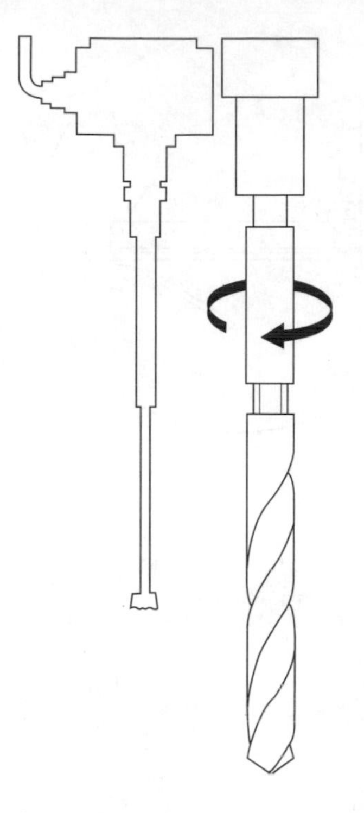

Figure T.1 Telescopic trip device for a drilling machine. Source: BS 5304 (1988) Safeguarding of Machinery.

tunnel guard A form of distance guard whereby access to the danger area of the machine is prevented in that materials must be passed into the machine via a metal enclosure or tunnel. Commonly used on metal-cutting machines, this form of guard incorporates a mechanical interlock whereby, if the tunnel guard is raised, the machine stops. See Figure T.2.

two-factor theory One of the most prominent theories of motivation developed by Herzberg (1957), which brought in the concept of job enrichment. Herzberg said: 'Before you can improve satisfaction with work, you must reduce dissatisfaction'.
Herzberg examined homeostatic needs of workers (hygiene factors or maintenance factors), which are concerned with avoiding discomfort, and the growth needs (motivators), which are concerned with the achievement of fulfilment and satisfaction.
The hygiene factors and motivators are outlined below:

Hygiene factors	Motivators
Money	Challenge
Working conditions	Responsibility
Safety arrangements	Advancement
Quality of supervision	Interest and stimulation created by the job

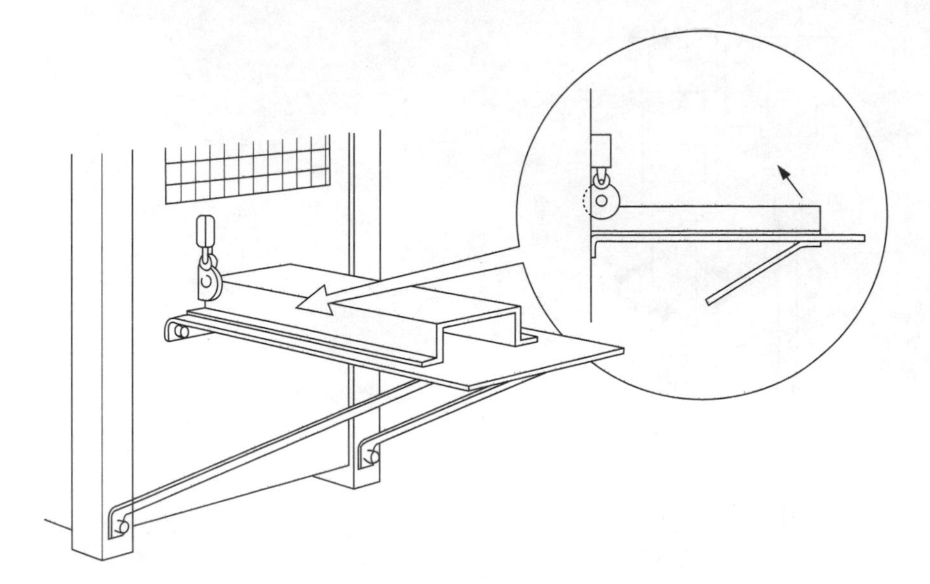

Figure T.2 Tunnel guard for a metal-cutting machine.

Administrative procedures	Achievement
Interpersonal relationships	Recognition
Status	Possibility of growth
Security	

He established that there was no one factor that determined the presence or absence of job satisfaction. However, he did establish that before satisfaction with the work situation can be improved, the factors that cause dissatisfaction must be recognized and measures implemented to remove this dissatisfaction.

two-hand control device A device which requires both hands to operate the machinery controls, thus affording a measure of protection from danger only to the machinery operator and not other persons. See Figure T.3.

Tyndall beam (1E2) This device enables respiratory range dust and fumes, which are commonly invisible to the naked eye in normal lighting, to be observed. The technique enables the identification of the source of dust in a process, and its direction and rate of movement, and allows a broad assessment to be made as to whether further, more sophisticated, techniques of assessment are required.
The lamp emits a horizontal beam of light, and when the beam passes through a dust cloud, forward scatter of light occurs which is visible to an observer looking along the beam in the direction of the lamp.

type 'E' equipment (2C2) In the case of flameproof electrical equipment, this includes equipment such as transformers and squirrel cage motors. This type of equipment does not normally produce sparks and is suitable for zone 2 conditions. See ***zoning***.

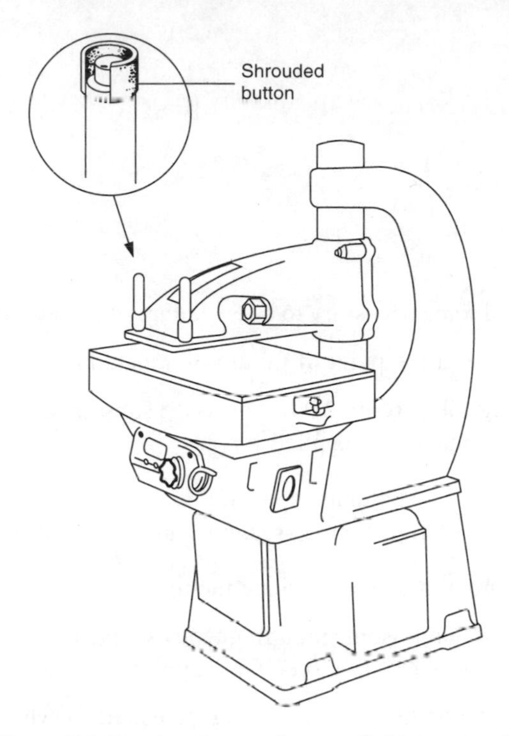

Figure T.3 Two-hand control on a clicking press. Source: BS 5304 (1988) Safeguarding of Machinery.

type 'N' equipment (2C2) A type of flameproof electrical equipment designed for use in zone 2 conditions so constructed that, properly used, it will not ignite flammable atmospheres under normal conditions. See ***zoning***.

U

ultimate load The minimum applied force necessary to cause failure of a material.

ultimate stress The intensity of stress at the point of failure of a material.

ultimate tensile strength (ultimate tensile stress) (2D4) This is the stress produced in a test piece of a material by the maximum applied force.

$$\text{Ultimate tensile strength} = \frac{\text{Maximum applied force}}{\text{Original cross-sectional area}}$$

Ultimate tensile strength is measured in newtons/square metre.

ultrasonic device A safety device using ultrasonic high frequency sound designed to detect the presence of a person or object in a specific danger area.

ultrasonic testing (2D4) A form of non-destructive testing of materials which entails transmitting waves of certain frequencies and observing reflections which indicate flaws and defects in the material.

ultraviolet analysis A technique for analysing and measuring concentrations of contaminants using radiation in the ultraviolet/visible radiation wavelength. The level of absorption of radiation at wavelengths specific to a particular contaminant is directly proportional to the concentration of the contaminant.

ultraviolet radiation A form of non-ionizing radiation produced particularly in arc welding activities.

unconfined vapour cloud explosion An explosion which results from the ignition, in the open air, of a cloud comprising a mixture of a flammable gas or vapour with air.

unconscious motivation Motivation that can be inferred from a person's behaviour, but the person does not realize the presence of the motive.

understanding This is a critical feature of the communication and learning processes. In the process of improving understanding, use of the FIDO principle, with respect to the process of learning, should be considered in terms of the messages that a trainer is sending out to trainees:

- F – Frequency.
- I – Intensity.
- D – Duration.
- O – Over again.

unfair dismissal (2B4) This implies the dismissal of an employee which was unjust or unfair. A claim for unfair dismissal is based on whether the employer's decision to dismiss can be justified by the reasons given for the decision and the way in which the dismissal was dealt with.
For a dismissal to be 'fair', it must satisfy a two-stage test:

1. It can be specified that the reason for dismissal is one of the following:
   - capability or qualification of the employee;
   - conduct of the employee;
   - redundancy;
   - where continued employment would contravene some other statute or regulation; or
   - 'some other substantial reason'.
2. The employee acted unreasonably in the circumstances.

In cases where any of the above criteria have not been met by an employer, a tribunal may confirm 'unfair dismissal'.

unsealed source (radiation) These are sources of radiation which are not contained and can take many forms, such as gases, liquids and particulates. Unsealed sources are particularly dangerous, as entry into the body takes place quickly in many cases.

upper flammable limit (upper explosive limit) (2C1) This is the highest concentration of a flammable gas or vapour which, when mixed with air, is capable of ignition and subsequent flame propagation under prescribed test conditions.

use In relation to work equipment means any activity involving work equipment and includes starting, stopping, programming, setting, transporting, repairing, modifying, maintaining, servicing and cleaning.
(Provision and Use of Work Equipment Regulations 1998)
In relation to display screen equipment, means use for or in connection with work.
[Health and Safety (Display Screen Equipment) Regulations 1992]

user (display screen equipment) An employee who habitually uses display screen equipment as a significant part of his normal work.
[Health and Safety (Display Screen Equipment) Regulations 1992]

V

value expressive function An important element of attitude, this function is used by people to express to the outside world a picture of themselves that is pleasing and satisfying to them. It is directly connected with self-image.

vaporizing liquid appliance This type of fire appliance consists of a cylinder containing the liquid which is pressurized to 10 bar with dry carbon dioxide or nitrogen. Striking the knob allows the pressure to expel the liquid.
This type of appliance should not be used in confined spaces as the liquids and combustion products are toxic.
This form of appliance is now obsolete.

vapour A gas which can be liquefied by the application of pressure alone without reduction in temperature.
This is the gaseous form of a material normally encountered in a solid or liquid state at normal temperature and pressure.

vapour pressure (2C1) The percentage gas by volume in air. See Figure V.1.

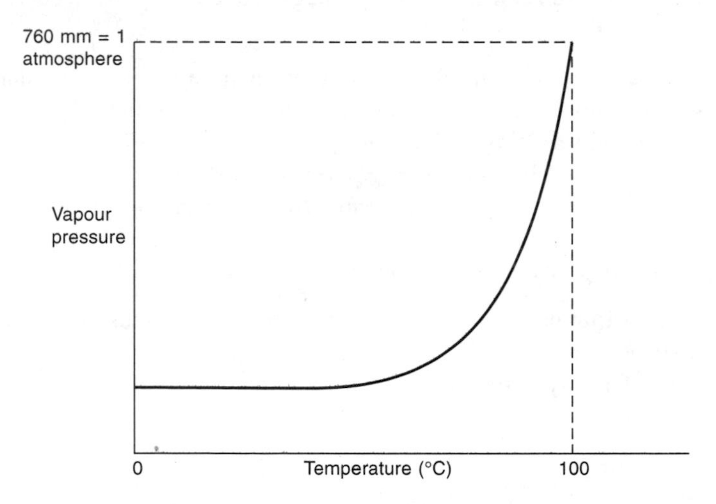

Figure V.1 The relationship between vapour pressure and temperature.

variable sequence robot This is similar to a fixed sequence robot other than the fact that set information can be more readily changed. This type of robot must be able to fulfil a range of movements in order to accommodate different sets or groups or instructions.

ventilation (1C1) This implies the movement of air through a building, which may be by natural or mechanical means, or both.
The principal requirements for a ventilation system are:

- the provision and maintenance of the circulation of fresh air in every occupied part of a workplace (comfort ventilation);
- the rendering harmless of all potentially injurious airborne contaminants, e.g. dust, fumes, gases (exhaust ventilation).

very toxic Substances and preparations which in very low quantities can cause death or acute or chronic damage to health when inhaled, swallowed or absorbed via the skin.
[Chemicals (Hazard Information and Packaging for Supply) Regulations 1994]

vessel A simple pressure vessel, being a welded vessel intended to contain air or nitrogen at a gauge pressure greater than 0.5 bar, not intended for exposure to flame, and having the following characteristics:

- The components and assemblies contributing to the strength of the vessel under pressure are made either of non-alloy quality steel, or of non-alloy aluminium, or of non-age hardening aluminium alloy.
- The vessel consists either:
  - of a cylindrical component with a circular cross-section, closed at each end, being outwardly dished or flat and also being co-axial with the cylindrical component;
  - of two co-axial outwardly dished ends.
- The maximum working pressure (PS) is not more than 30 bar, and the PS.V not more than 10 000 bar litres.
- The minimum working temperature is not lower than –50°C and the maximum working temperature is not higher than 300°C in the case of steel vessels and 100°C in the case of aluminium or aluminium alloy vessels.

[Simple Pressure Vessels (Safety) Regulations 1991]

vibration (1E1) A body is said to vibrate when it describes an oscillating motion about a fixed position. Vibration is measured on the basis of frequency, which is the number of times a complete motion cycle takes place during the period of 1 second, and expressed in hertz (Hz). See Figure V.2.

vicarious liability (1B1, 2B3) This form of liability is based on the fact that if an employee, whilst acting in the course of his or her employment, negligently injures another employee or the employee of another employer, his or her employer, rather than the employee, will be liable for that injury.
Vicarious liability rests on the employer simply as a result of the fact that he or she is the employer and is deemed to have ultimate control over his or her employees, i.e. the Master and Servant relationship.
This liability must be insured against under the Employers' Liability (Compulsory Insurance) Act 1969. Employers cannot contract out of this liability as it is

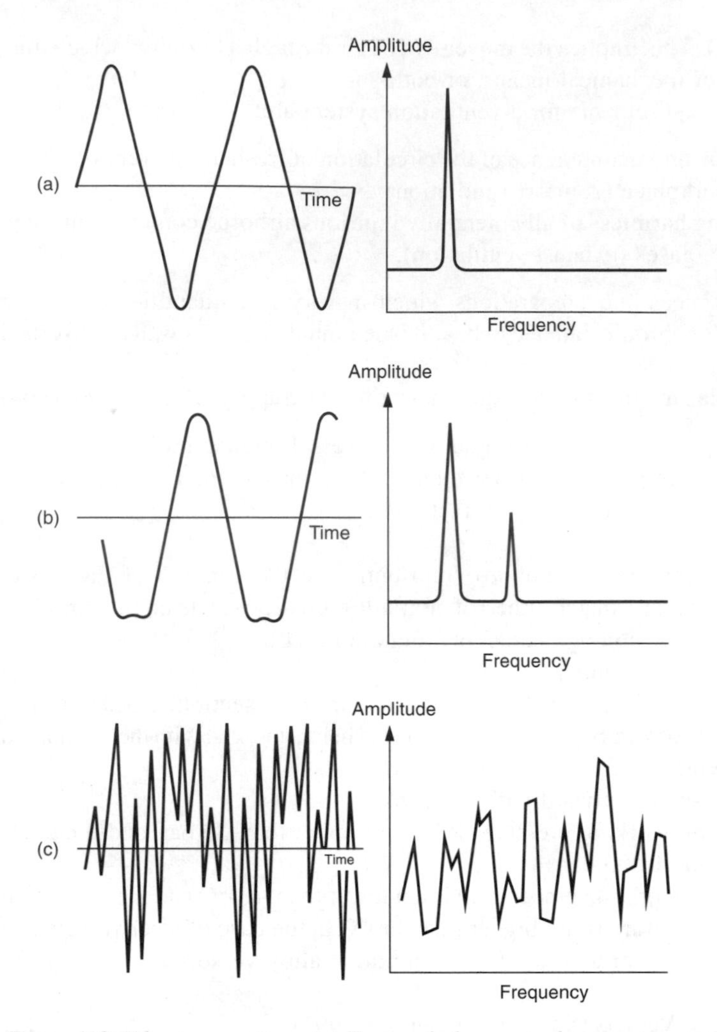

Figure V.2 Vibratory motion. (a) Tuning fork – a single component vibrating at a single frequency. (b) Internal combustion engine – several components vibrating at different frequencies simultaneously. (c) Factory machinery – a large number of components vibrating at different frequencies simultaneously.

prohibited by the Law Reform (Personal Injuries) Act 1948 and the Unfair Contract Terms Act 1977.

visible light A form of non-ionizing radiation produced by intense sources of light, such as the sun, arc lamps and electric welding operations, which can damage the eye.

visual fatigue (2E3) This is associated with eye strain and discomfort experienced by many people using display screen equipment, which may be due to a combination of both poor workstation design and eyesight defects. Vision screening of display screen equipment users should be undertaken on a regular basis and where users are experiencing symptoms of visual fatigue.

volenti non fit injuria A Latin term meaning 'to one who is willing no harm is done', more commonly referred to as 'voluntary assumption of risk'. As such, it is a complete defence and no damages would be payable to a claimant. However, it is a very limited defence based on the employee voluntarily assuming a risk as part of his or her normal work activities.
[Smith v. Baker & Sons (1891) AC 305]

W

warrant A written authority issued by a magistrate, addressed to a constable directing him or her to carry out some specified act, namely to arrest the person named in the warrant, and to bring that person before the court.

wastage A form of boiler corrosion resulting in loss of metal thickness in, for instance, the external shell crown, firebox, cross tube and around boiler mountings, with a resulting loss in metallic strength.

waste (2C6) (a) Any substance which constitutes a scrap material or effluent or other unwanted surplus substance arising from the application of any process. (b) Any substance and article which requires to be disposed of as being broken, worn out, contaminated or otherwise spoiled.
(Environmental Protection Act 1990)
Waste may take many forms:

- solid, liquid or gas;
- flammable or non-flammable;
- biodegradable;
- clinical;
- radioactive;
- hazardous or safe;
- special.

waste disposal (2C6) The process of disposing of waste by a variety of processes:

- landfill;
- land raising;
- total incineration;
- selective incineration;
- physical and chemical treatment processes;
- separation of reusable or recyclable items.

Waste Management Licence (2C6) A licence granted by a Waste Regulation Authority authorizing the treatment, keeping or disposal of any specified description of controlled waste by means of a specified mobile plant.
(Environmental Protection Act 1990)

water appliance There are three types of water-containing fire appliance: soda acid, gas cartridge and stored pressure. See individual entries.

water gauge A tubular gauge situated so as to indicate the level of water in a steam boiler. Generally, steam boilers with an evaporative capacity exceeding 136 kg of steam per hour should be fitted with two water gauges.

wear (2D4) This is essentially a deterioration in the surface finish of a material caused by breaking down of a surface coating or finish. It may be caused by:

- Scuffing – whereby moving parts of a structure rub together.
- Pitting – a form of surface fatigue failure arising from the formation of minute pits in the surface.
- Abrasion – caused by foreign particles deposited between moving surfaces which are harder than the moving surfaces.
- Fretting – occurring in pin-jointed components, where there should be no motion, but where there is a small amount of motion between joints and parts. Fretting destroys the surface finish, which promotes metal fatigue and corrosion.

weighting networks These networks are incorporated in a sound pressure level meter providing a range of sensitivities to sounds of differing frequencies. The principal objective is to simulate the response characteristics of the human ear at different frequencies.

Weighting networks include the A, B, C and D scales, which are used for different aspects of noise measurement. The A scale is normally used for industrial noise measurement as it is the scale which most closely corresponds to the performance of the human ear.

welfare amenity provisions (1C1) Detailed requirements relating to the provision and maintenance of welfare amenity provisions are covered in the Workplace (Health, Safety and Welfare) Regulations 1992 and accompanying Approved Code of Practice.

Duties on employers relating to welfare amenities include the provision of:

- suitable and sufficient sanitary conveniences and washing facilities;
- an adequate supply of wholesome drinking water;
- accommodation for clothing;
- facilities for changing clothing;
- facilities for rest and the taking of meals.

Schedule 1 to the regulations covers provisions with regard to the number of sanitary fitments and washing facilities to be provided.

wet bulb globe temperature (1E2, 2E3) In the assessment of thermal stress, it is common to use an Index of Thermal Stress. The most commonly used index is that based on physiological observations and related to wet bulb globe temperature (WBGT) shown with a whirling hygrometer.

The following equations are used to calculate WBGT:

Outdoor work with a solar load:

$$WBGT = 0.7WB + 0.2GT + 0.1DB$$

Indoor work, or outdoor work with no solar load:

$$WBGT = 0.7WB + 0.3GT$$

where WB = natural wet bulb temperature, DB = dry bulb temperature and GT = globe thermometer temperature.

The natural wet bulb temperature is that recorded from the sling hygrometer without any rotation of the sling.

Wiring Regulations (2C2) See ***IEE Regulations***.

work Work as an employee or as a self-employed person.
(Health and Safety at Work etc. Act 1974)

work equipment (1D1) This is defined as including any machinery, appliance, apparatus or tool and any assembly of components that, in order to achieve a common end, are arranged and controlled so that they function as a whole.
(Provision and Use of Work Equipment Regulations 1998)

Working Group on the Assessment of Toxic Chemicals (WATCH) (1E2) The organization responsible for approving and reviewing occupational exposure limits in the United Kingdom.

working platform A specifically designed working platform attached to the tines of a lift truck and used as a means of elevating people to undertake a range of tasks, e.g. picking goods from pallet racking systems, high level repairs and maintenance.

workplace Any premises or part of them that are not domestic premises and are made available to anyone as a place of work, including:

(a) any place within the premises to which such a person has access while at work;
(b) any room, lobby, corridor, staircase, road or other place used as a means of access to or egress from the workplace or where facilities are provided for use in connection with the workplace other than a public road.

[Workplace (Health, Safety and Welfare) Regulations 1992]

work-related upper limb disorders Work-related upper limb disorders caused by repetitive strain injuries (RSI) were first defined by Bernardo Ramazzini, the Italian father of occupational medicine, in the early eighteenth century.
The International Labour Organization recognized RSI as an occupational disease in 1960, defining it as ‘a condition caused by forceful, frequent, twisting and repetitive movements’.
RSI can take a number of forms:

- epicondylitis – inflammation of the area where a muscle joins a bone;
- peritendinitis – inflammation of the area where a tendon joins a muscle;
- carpal tunnel syndrome – a painful condition in the area where nerves and

tendons pass through thc carpal bone in the hand;
- tenosynovitis – inflammation of the synovial lining of the tendon sheath;
- tendinitis – inflammation of the tendons, particularly in the fingers;
- Dupuytren's Contracture – a condition affecting the palm of the hand, where it is impossible to straighten the hand and fingers;
- writer's cramp – a condition causing cramps in the hand, forearm and fingers.

workstation (1A2) This means an assembly comprising:

(a) display screen equipment (whether provided with software determining the interface between the equipment and its operator user, a keyboard or any other input device);
(b) any optional accessories to the display screen equipment;
(c) any disk drive, telephone, modem, printer, document holder, work chair, work desk, work surface or other item peripheral to the display screen equipment; and
(d) the immediate environment around the display screen equipment.

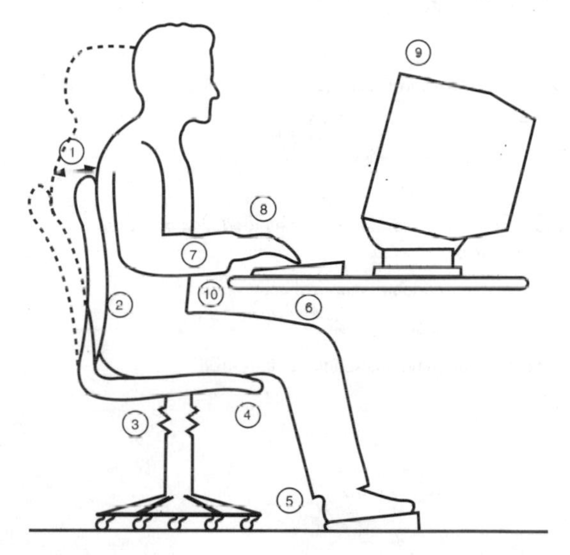

(1) Seat back adjustability
(2) Good lumbar support
(3) Seat height adjustability
(4) No excess pressure on underside of thighs and back of knees
(5) Foot support if needed
(6) Space for postural change, no obstacles under desk
(7) Forearms approximately horizontal
(8) Minimal extension, flexion or deviation of wrists
(9) Screen height and angle should allow comfortable head position
(10) Space in front of keyboard to support hands/wrists during pauses in keying

Figure W.1 Seating and posture for typical office tasks.

[Health and Safety (Display Screen Equipment) Regulations 1992]
See Figures W.1 and W.2.

(1) Adequate lighting

(2) Adequate contrast, no glare or distracting reflections

(3) Distracting noise minimized

(4) Leg room clearances to allow postural changes

(5) Window covering

(6) Software: appropriate to task, adapted to user, provides feedback on system status, no undisclosed monitoring

(7) Screen: stable image, adjustable, readable, glare/reflection-free

(8) Keyboard: usable, adjustable, detachable, legible

(9) Work surface: allows flexible arrangements, spacious, flare-free

(10) Work chair: adjustable

(11) Footrest

Figure W.2 DSE workstation design and layout.

XYZ

X-rays A form of ionizing radiation emitted from metals when bombarded with high-energy electrons. X-rays are produced by changes in the state of planetary electrons and are propagated as a wave.

yield point (2D4) A stage in a tensile test of a metal where the elastic extension stage terminates and the metal passes into the plastic stage, at which point the metal will not return to its original shape.

young persons A young person is someone who has ceased to be a child, but who is not yet 18 years old.
Under the Health and Safety (Youth Training Scheme) Regulations 1983, all young people on youth training schemes are covered, as regards their health and safety at work, in the same way as other employees over the age of 18 years.
Under the Management of Health and Safety at Work Regulations 1999, employers are required to:

- take particular account of young workers' lack of experience, absence of awareness of existing or potential risks, or their immaturity, when they assess the risks to their health and safety;
- the assessment must be made before the young person begins work and must address specific factors laid out in the regulations;
- take account of the risk assessment in determining whether the young person is prohibited from doing certain work;
- the requirement does not apply to young people over school-leaving age where the work is necessary for training, they are properly supervised and the risks are reduced to the lowest practicable level;
- inform parents, or those with parental responsibility for children, of the outcome of the risk assessment and the control measures introduced.

Young's Modulus (Modulus of Elasticity) This is expressed as a constant:

$$E = \frac{\text{Stress}}{\text{Strain}}$$

This constant is different for different materials and is the force required to stretch a bar of the material of unit sectional area to twice its original length, assuming the material to be perfectly elastic.

zoning (flammable atmospheres) (2C2) The use of electrical equipment in flammable atmospheres can be a substantial risk.

On this basis, zones where flammable atmospheres can arise are classified thus:

- Zone 0 – a zone in which a flammable atmosphere is known to be continuously present, or present for long periods.
- Zone 1 – a zone in which a flammable atmosphere is likely to occur, at least during normal working.
- Zone 2 – a zone in which a flammable atmosphere is unlikely to occur save under abnormal conditions, such an occurrence being of only short duration.

zoonoses (2E4) The group of diseases transmissible from animals to Man, and including brucellosis, leptospirosis and anthrax.

Bibliography and Further Reading

Allport, G.W. (1961). *Pattern and Growth in Personality.* New York: Holt, Rinehart and Winston.

Bass, B.M. (1965). *Organisational Psychology*. Boston: Allyn and Bacon.

Bird, F.E. and Loftus, R.G. (1984). *Loss Control Management*. Birmingham: RoSPA.

British Standards Institution (1975). BS 5304, *Code of Practice: Safeguarding of Machinery*. Milton Keynes: BSI.

Brown, J.A.C. (1972). *The Social Psychology of Industry*. Pelican Books.

Bruel & Kjaer Instruments Inc. (1988). *Measuring Sound*. Naerum, Denmark: Bruel & Kjaer Instruments Inc.

Chemical Industries Association (1977). *A Guide to Hazard and Operability Studies*. London: Chemical Industries Association.

Confederation of British Industries (1991). *Developing a Safety Culture*. London: CBI.

Cooper, C.L., Cooper, R.D. and Eaker, I.H. (1988). *Living with Stress*. Penguin.

Cooper, C.L. and Marshall, J. (1978). *Organisational Stress for the Individual Manager.*

Department of Employment (1973). *Safety Training Needs and Facilities in One Industry*. London: HMSO.

Department of Employment (1998). *The Working Time Regulations 1998*. London: HMSO.

Department of Employment and Productivity (1978). *Glossary of Training Terms*. London: HMSO.

Department of Health and Social Security (1975). *The Social Security Act 1975*. London: HMSO.

Department of Trade and Industry (1978). *The Consumer Protection Act 1978*. London: HMSO.

Dewis, M. and Stranks, J. (1990). *Health and Safety at Work Handbook*. Croydon: Tolley.

Fletcher, J.A. and Douglas, H.M. (1971). *Total Loss Control*. London: Associated Business Programmes.

Hale, A.R. and Hale, M. (1970). Accidents in perspective. *Occupational Psychology* **44**: 115–21.

Health and Safety Commission (1976). *Factories Act 1961: A Short Guide*. London: HMSO.

Health and Safety Commission (1987). *Human Reliability Assessment – A Critical Overview*. Advisory Committee on the Safety of Nuclear Installations. London: HMSO.

Health and Safety Commission (1988). *Safety Representatives and Safety Committees*. London: HMSO.

Health and Safety Commission (1990). *First Aid at Work: Health and Safety (First Aid) Regulations 1981 and Guidance, Approved Code of Practice*. London: HMSO.

Health and Safety Commission (1994). *Managing Construction for Health and Safety: Construction (Design and Management) Regulations 1994, Approved Code of Practice*. London: HMSO.

Health and Safety Commission (1998). *Safe Use of Work Equipment: Provision and Use of Work Equipment Regulations 1998, Approved Code of Practice and Guidance*. London: HMSO.

Health and Safety Commission (2000). *Health and Safety in Small Firms*. Sudbury: HSE Books.
Health and Safety Executive (1980). *Flame Arrestors and Explosion Reliefs*. London: HMSO.
Health and Safety Executive (1980). *Safe Working with Lift Trucks*. Guidance Note G 6. London: HMSO.
Health and Safety Executive (1987). *Introduction to Local Exhaust Ventilation*. Guidance Note G 37. London: HMSO.
Health and Safety Executive (1987). *Lighting at Work*. London: HMSO.
Health and Safety Executive (1988). *Control of Asbestos at Work: The Control of Asbestos at Work Regulations 1987*. London: HMSO.
Health and Safety Executive (1988). *Rider Operated Lift Trucks – Operator Training: Approved Code of Practice and Supplementary Guidance*. London: HMSO.
Health and Safety Executive (1989). *Construction (Head Protection) Regulations 1989: Guidance on the Regulations*. London: HMSO.
Health and Safety Executive (1989). *Essentials of Health and Safety at Work*. London: HMSO.
Health and Safety Executive (1989). *Human Factors in Industrial Safety*. Guidance Note G 48. London: HMSO.
Health and Safety Executive (1989). *Memorandum of Guidance on the Electricity at Work Regulations 1989: Guidance on the Regulations*. London, Sheffield: HSE Enquiry Points.
Health and Safety Executive (1989). *Our Health and Safety Policy Statement: Writing your Health and Safety Policy Statement; Guide to Preparing a Safety Policy Statement for a Small Business*. London: HMSO.
Health and Safety Executive (1989). *Safe Systems of Work*. London and Sheffield: HSE Enquiry Points.
Health and Safety Executive (1989, 1990). *Noise at Work: Noise Guides Nos 1 to 8, Noise at Work Regulations 1989*. London: HMSO.
Health and Safety Executive (1990). *A Guide to the Health and Safety at Work etc. Act 1974: Guidance on the Act*. London: HMSO.
Health and Safety Executive (1991). *Successful Health and Safety Management*. London: HMSO.
Health and Safety Executive (1992). *Display Screen Equipment: Guidance on the Health and Safety (Display Screen Equipment) Regulations 1992*. London: HMSO.
Health and Safety Executive (1992). *Listen Up!* Sudbury: HSE Books.
Health and Safety Executive (1992). *Manual Handling: Guidance on the Manual Handling Operations Regulations 1992*. London: HMSO.
Health and Safety Executive (1992). *Personal Protective Equipment at Work: Guidance on the Personal Protective Equipment at Work Regulations 1992*. London: HMSO.
Health and Safety Executive (1992). *The Noise at Work Regulations 1989: A Brief Guide to the Requirements for Controlling Noise at Work*. Sudbury: HSE Books.
Health and Safety Executive (1992). *Workplace Health, Safety and Welfare: Approved Code of Practice, Workplace (Health, Safety and Welfare) Regulations 1992*. London: HMSO.
Health and Safety Executive (1994). *Successful Health and Safety Management*. Sudbury: HSE Books.
Health and Safety Executive (1994). *Upper Limb Disorders: Assessing the Risks*. Sudbury: HSE Books.
Health and Safety Executive (1995). *Managing Vehicle Safety at the Workplace*. Sudbury: HSE Books.
Health and Safety Executive (1995). *Stress at Work*. Sudbury: HSE Books.
Health and Safety Executive (1995). *The Complete Idiot's Guide to CHIP2: Chemicals (Hazard Information and Packaging for Supply) Regulations 1994*. Sudbury: HSE Books.

Health and Safety Executive (1996). *A Guide to the Construction (Health, Safety and Welfare) Regulations 1996*. Sudbury: HSE Books.
Health and Safety Executive (1996). *Consulting Employees on Health and Safety: A Guide to the Law*. Sudbury: HSE Books.
Health and Safety Executive (1996). *Electrical Safety and You*. Sudbury: HSE Books.
Health and Safety Executive (1996). *Getting to Grips with Manual Handling: A Short Guide for Employers*. Sudbury: HSE Books.
Health and Safety Executive (1996). *Health and Safety in Construction*. Guidance Note G 150. Sudbury: HSE Books.
Health and Safety Executive (1996). *Preventing Dermatitis at Work: Advice for Employers and Employees*. Sudbury: HSE Books.
Health and Safety Executive (1996). *Protecting your Health at Work*. Sudbury: HSE Books.
Health and Safety Executive (1996). *Signpost to the Health and Safety (Safety Signs and Signals) Regulations 1996*. Sudbury: HSE Books.
Health and Safety Executive (1996). *The Reporting of Injuries, Diseases and Dangerous Occurrences*. London: HMSO.
Health and Safety Executive (1997). *Computer Control: A Question of Safety*. Sudbury: HSE Books.
Health and Safety Executive (1998). *Buying New Machinery*. Sudbury: HSE Books.
Health and Safety Executive (1999). *Violence at Work: A Guide for Employers*. Sudbury: HSE Books.
Health and Safety Executive (1999). *Workplace Health, Safety and Welfare*. Sudbury: HSE Books.
Health and Safety Executive (2000). *Occupational Exposure Limits*. Guidance Note EH 40. London: HMSO.
Heinrich, H.W. (1931). *Unsafe Acts and Conditions*. London: McGraw-Hill.
Henderson, J. (1992). *A Guide to the Employment Acts*. London: Nicholas Brealey.
Herzberg, F. (1957). *The Two Factory Theory*. New York: McGraw-Hill.
Herzberg, F., Mansner, B. and Snyder, M. (1959). *The Motivation to Work*. New York: John Wiley.
Home Office (1987). *The Fire Safety and Safety of Places of Sport Act 1987*. London: HMSO.
Home Office and Scottish Home and Health Departments (1977). *Guides to the Fires Precautions Act 1971*. London: HMSO.
Institution of Electrical Engineers (1998). *IEE Regulations for Electrical Installations (The Wiring Regulations)*. Hitchin, Herts: IEE.
McGregor, D. (1960). *The Human Side of Enterprise*. New York: McGraw-Hill.
Maslow, A.H. (1954). *Motivation and Personality*. New York: Harper.
Morgan, C.T. and King, R.A. (1971). *Introduction to Psychology*. London: McGraw-Hill.
Rimington, J.R. (1989). *The Onshore Safety Regime: HSE Director General's Submission to the Piper Alpha Enquiry*. London: HMSO.
Secretary of State for Employment (1974). *Health and Safety at Work etc. Act 1974*. London: HMSO.
Secretary of State for Employment (1998). *The Lifting Operations and Lifting Equipment Regulations 1998*. London: HMSO.
Secretary of State for Employment (1998). *The Provision and Use of Work Equipment Regulations 1998*. London: HMSO.
Secretary of State for Employment (1999). *The Control of Substances Hazardous to Health Regulations 1999*. London: HMSO.
Secretary of State for Employment (1999). *The Management of Health and Safety at Work Regulations 1999 and Approved Code of Practice*. London: HMSO.

Secretary of State for the Environment (1990). *The Environmental Protection Act 1990*. London: HMSO.
Selye, H. (1936). *The Stress of Life*. Revised 1976. New York: McGraw-Hill.
Stranks, J. (2000). *The Handbook of Health and Safety Practice*. Pearson Education.
Stranks, J. (2001). *A Manager's Guide to Health and Safety at Work*. London: Kogan Page.
Stranks, J. (2001). *Health and Safety Law*. Pearson Education.
Stranks, J. (2001). *One Stop Health and Safety*. London: ICSA.
Street, A. and Alexander, W. (1972). *Metals in the Service of Man*. London: Penguin Books.